# NUMBERS AND C
## 1 THROUG

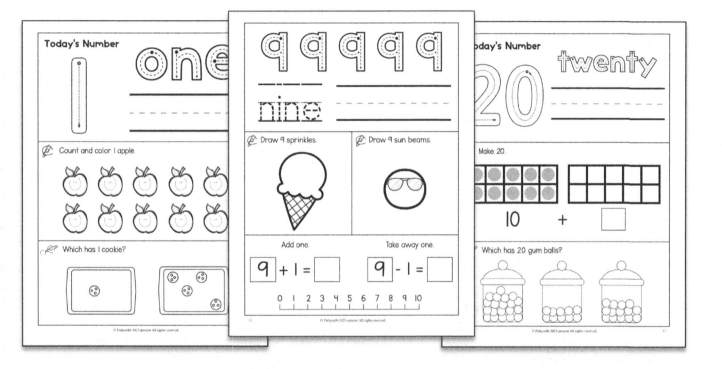

# KINDERGARTEN MATH

## Number Recognition
## Counting Objects
## Writing Numbers
## and Number Words
## 1 More and 1 Less

ISBN: 9798872798194

First printing edition 2023.

10 9 8 7 6 5 4 3 2 1

Published in Ormond Beach, Florida.

Contact the author :
fishyrobb.com

# Today's Number

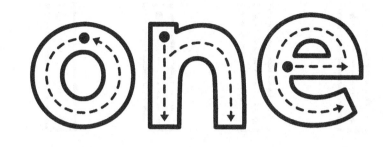

 Count and color 1 apple.

 Which has 1 cookie?

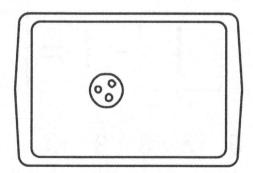

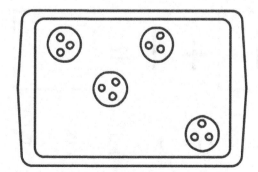

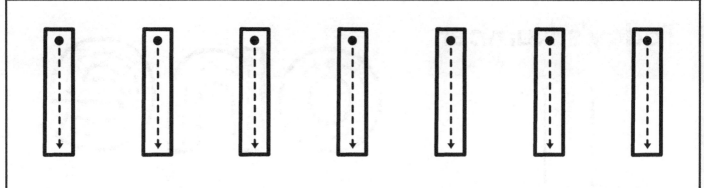

✎ Draw 1 flower.

✎ Draw 1 fish.

Add one.

Take away one.

$$1 + 1 = \boxed{\phantom{0}}$$

$$1 - 1 = \boxed{\phantom{0}}$$

0  1  2  3  4  5  6  7  8  9  10

# Count and color 1 bear.

# Find and color one.

| 1 | 2 | 3 | 4 | 5 | 6 | 7 | 8 | 9 | 10 |

# Draw 1 straw.

# Draw 1 cherry.

I less                                      I more

[ ]         [ 1 ]         [ ]

0  1  2  3  4  5  6  7  8  9  10

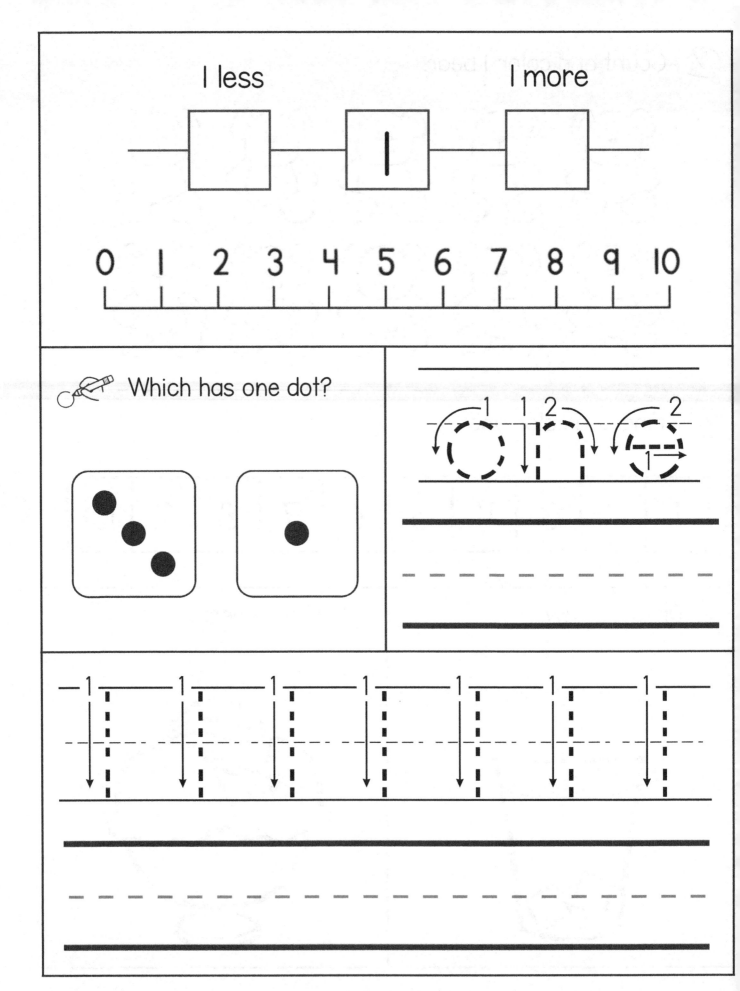

Which has one dot?

# Today's Number

 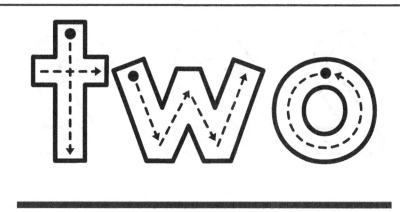

---

 Count and color 2 apples.

---

 Which has 2 fish?

two

 Draw 2 wheels.

 Draw 2 birds.

Add one.

$2 + 1 = \boxed{\phantom{0}}$

Take away one.

$2 - 1 = \boxed{\phantom{0}}$

0  1  2  3  4  5  6  7  8  9  10

Count and color 2 bears.

Find and color two.

| 1 | 2 | 3 | 4 | 5 | 6 | 7 | 8 | 9 | 10 |

Draw 2 leaves.

Draw 2 eggs.

I less                    I more

┌─────┐        ┌─────┐        ┌─────┐
│     │────────│  2  │────────│     │
└─────┘        └─────┘        └─────┘

0   1   2   3   4   5   6   7   8   9   10

✎ Which has two dots?

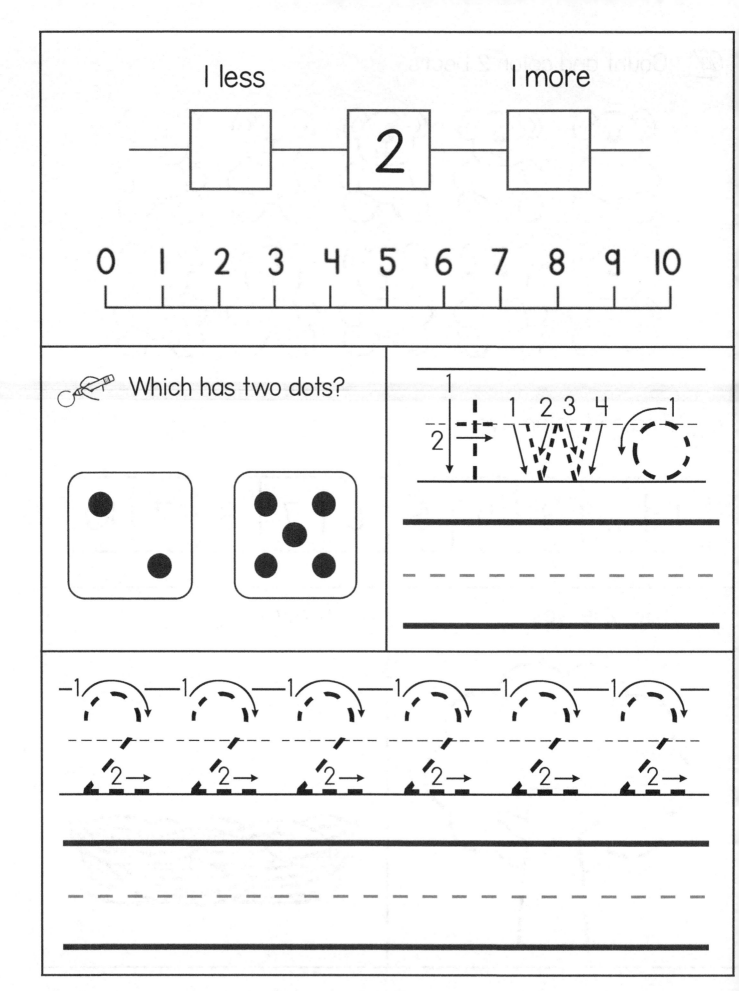

# Today's Number

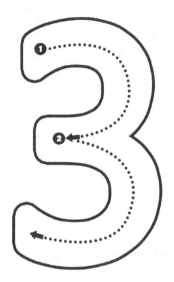

_____

- - - - - - - - - - - - - - - -

_____

 Count and color 3 apples.

 Which has 3 fingers?

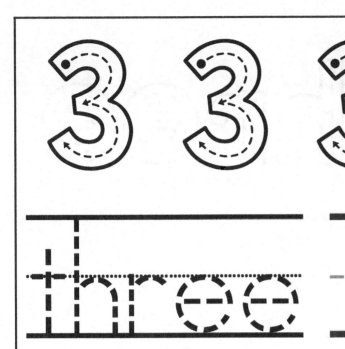

three

 Draw 3 scoops.

 Draw 3 candles.

Add one.

Take away one.

$3 + 1 = \boxed{\phantom{0}}$

$3 - 1 = \boxed{\phantom{0}}$

0  1  2  3  4  5  6  7  8  9  10

Count and color 3 bears.

Find and color three.

| 1 | 2 | 3 | 4 | 5 | 6 | 7 | 8 | 9 | 10 |

Draw 3 buttons.

Draw 3 ants.

I less          I more

☐   3   ☐

0 1 2 3 4 5 6 7 8 9 10

Which has three dots?

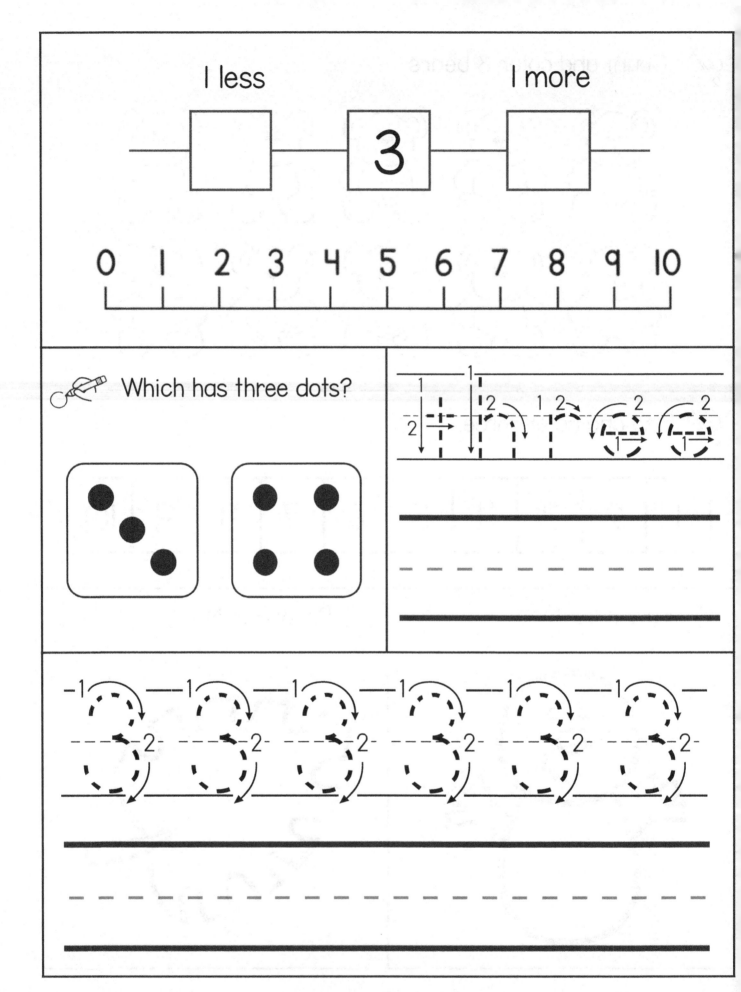

    

# Today's Number

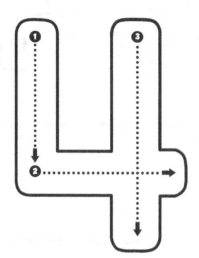
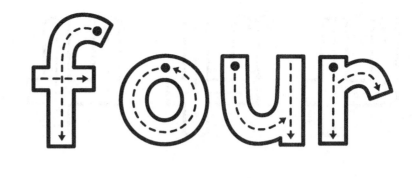

_____

- - - - - - - - - - -

_____

 Count and color 4 apples.

 Which has 4 marbles?

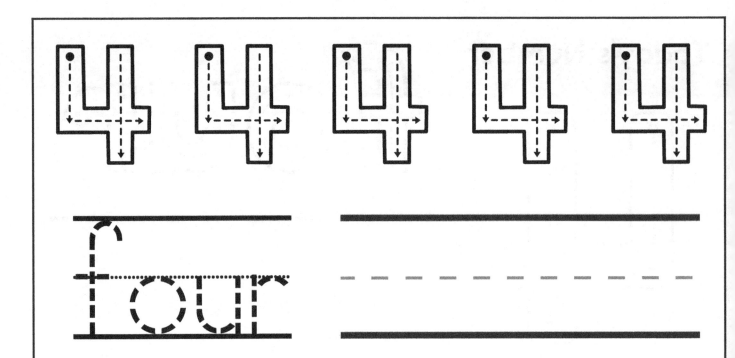

 Draw 4 frogs.

 Draw 4 petals.

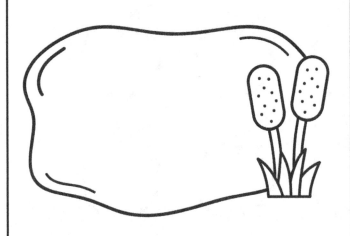

Add one.

Take away one.

| 4 | + 1 = | |

| 4 | - 1 = | |

0  1  2  3  4  5  6  7  8  9  10

✏️ Count and color 4 bears.

✏️ Find and color four.

| 1 | 2 | 3 | 4 | 5 | 6 | 7 | 8 | 9 | 10 |

✏️ Draw 4 fish.

✏️ Draw 4 apples.

I less                    I more

[ ]        4        [ ]

0  1  2  3  4  5  6  7  8  9  10

Which has three dots?

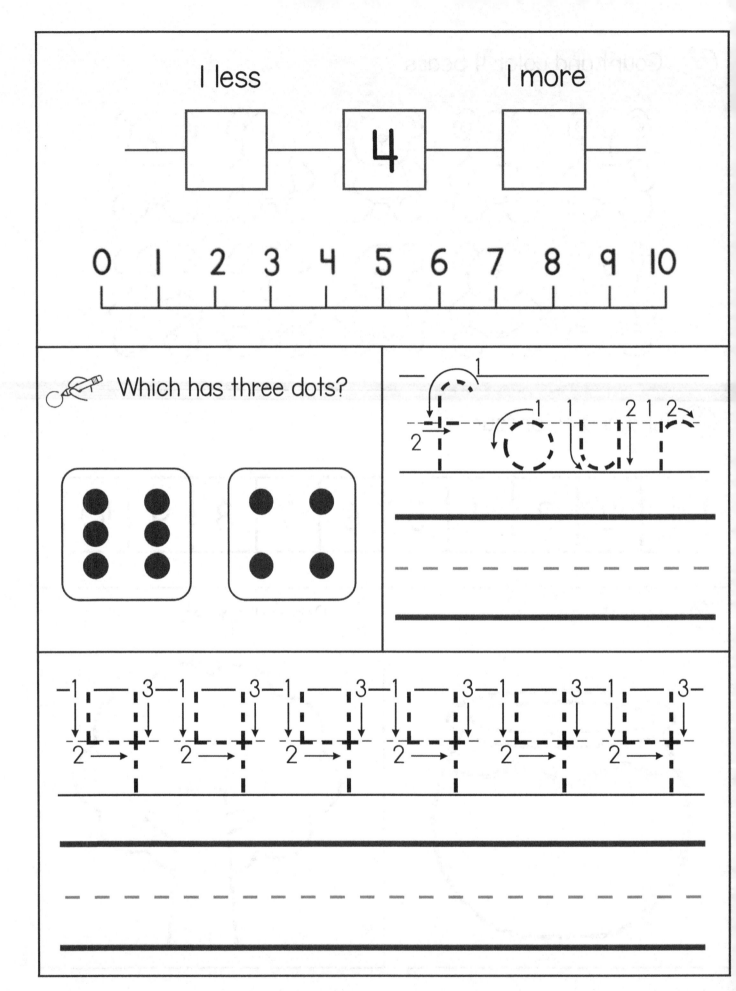

four

# Today's Number

 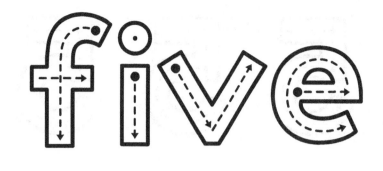

 Count and color 5 apples.

 Which has 5 fish?

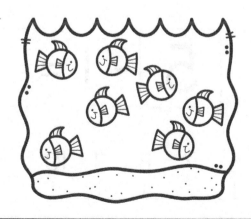 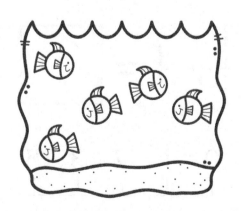

✐ Draw 5 bracelets.

Add one.

Take away one.

5 + 1 = ☐

5 - 1 = ☐

0  1  2  3  4  5  6  7  8  9  10

Count and color 5 bears.

Find and color five.

| 1 | 2 | 3 | 4 | 5 | 6 | 7 | 8 | 9 | 10 |
|---|---|---|---|---|---|---|---|---|----|

 Draw 5 pieces of popcorn.

Draw 5 eggs.

I less                          I more

[ ]          | 5 |          [ ]

0  1  2  3  4  5  6  7  8  9  10

Which has five dots?

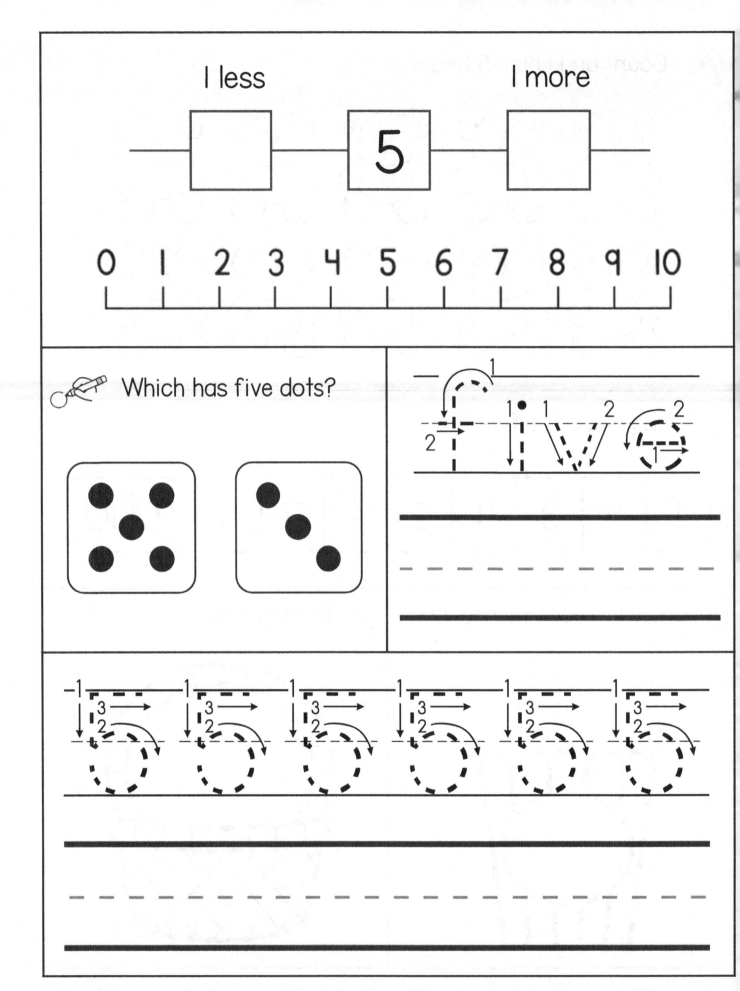

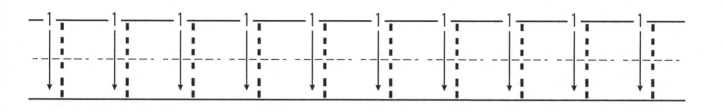

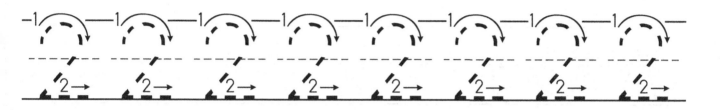

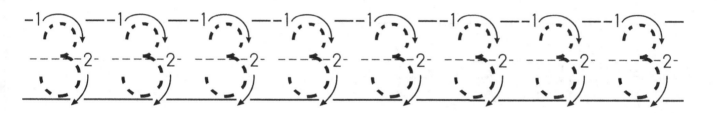

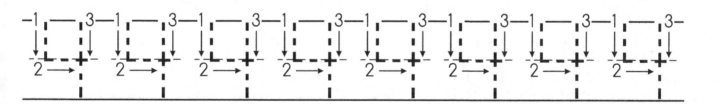

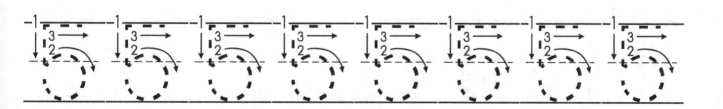

23

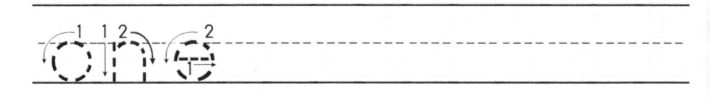

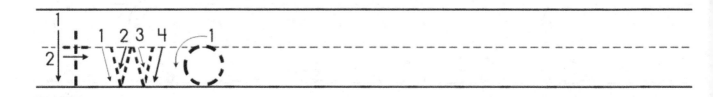

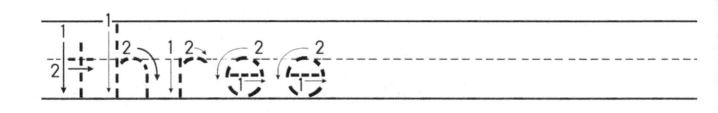

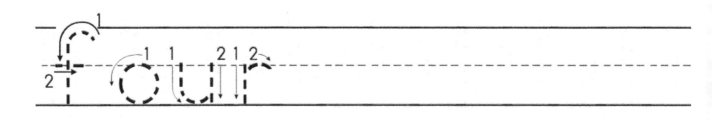

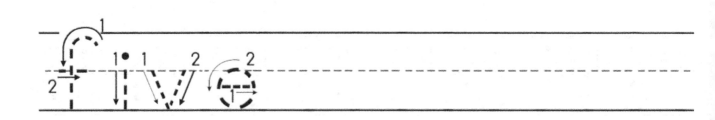

# Today's Number

 Count and color 6 apples.

 Which has 6 apples?

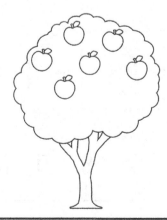 

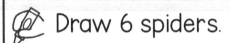 

six

 Draw 6 spiders.

✎ Draw 6 cars.

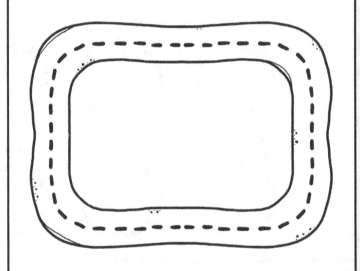

Add one.

Take away one.

6 + 1 = ☐

6 - 1 = ☐

0  1  2  3  4  5  6  7  8  9  10

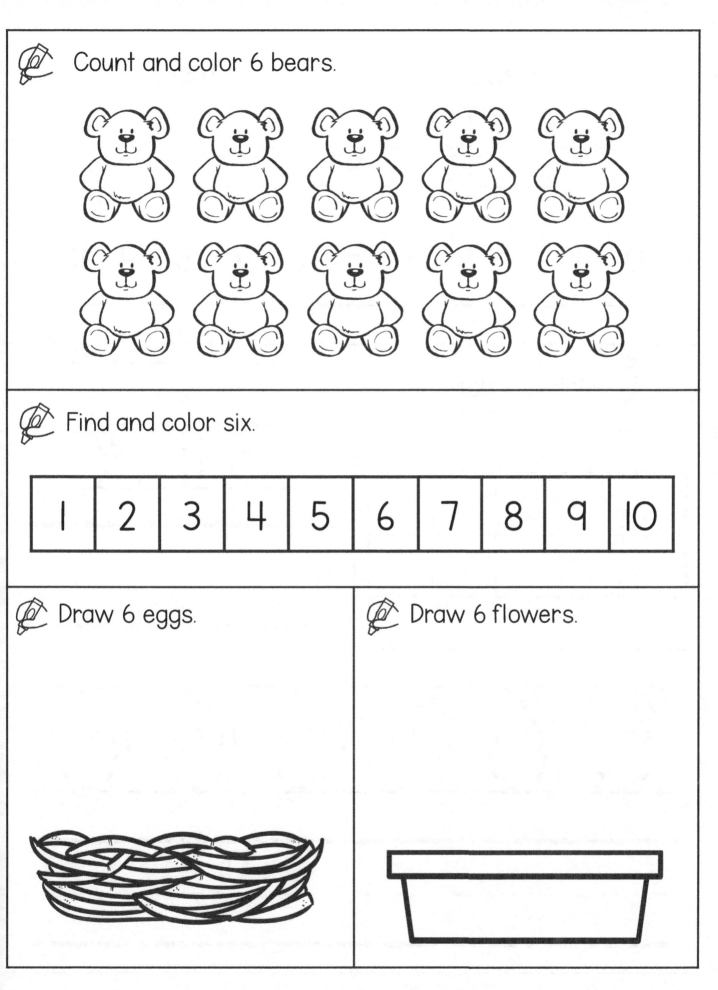

Count and color 6 bears.

Find and color six.

| 1 | 2 | 3 | 4 | 5 | 6 | 7 | 8 | 9 | 10 |

Draw 6 eggs.

Draw 6 flowers.

I less                    I more

[ ]        **6**        [ ]

0  1  2  3  4  5  6  7  8  9  10

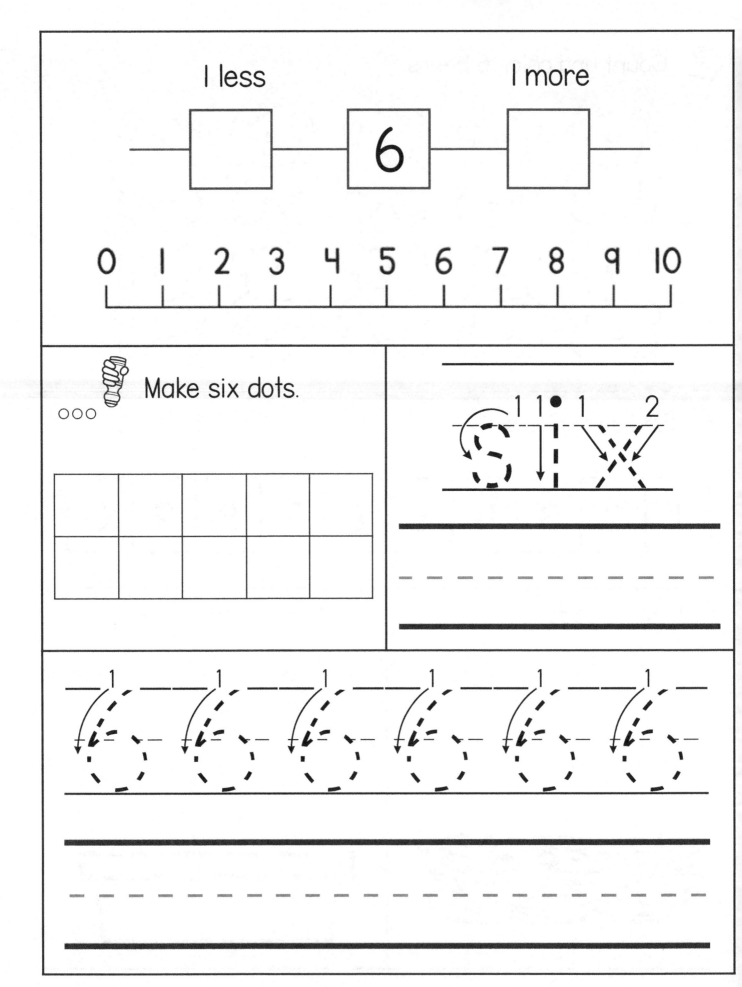

Make six dots.

six

# Today's Number

---

Count and color 7 apples.

---

Which has 7 cookies?

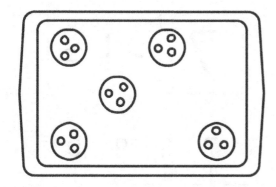 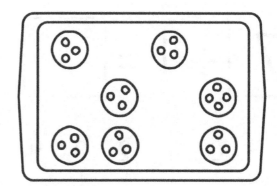

_____ _____

seven _____

 Draw 7 candies.

 Draw 7 jellyfish.

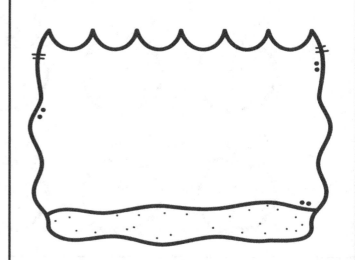

Add one.

$\boxed{7} + 1 = \boxed{\phantom{0}}$

Take away one.

$\boxed{7} - 1 = \boxed{\phantom{0}}$

0  1  2  3  4  5  6  7  8  9  10

Count and color 7 bears.

Find and color seven.

| 1 | 2 | 3 | 4 | 5 | 6 | 7 | 8 | 9 | 10 |

Draw 7 apples.

Draw 7 worms.

I less                          I more

[ ]          7          [ ]

0  1  2  3  4  5  6  7  8  9  10

Make seven dots.

seven

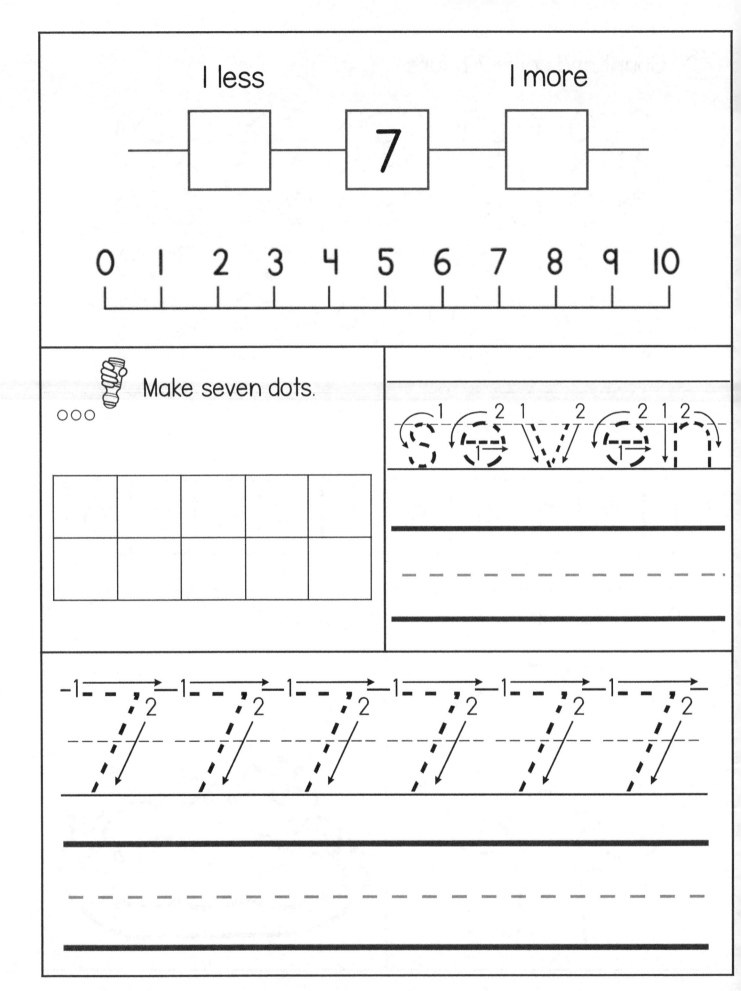

# Today's Number

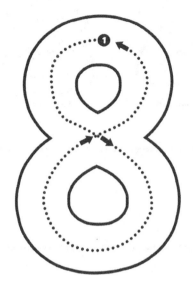

 Count and color 8 apples.

 Which has 8 ants?

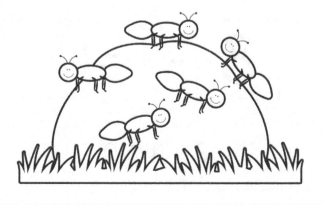
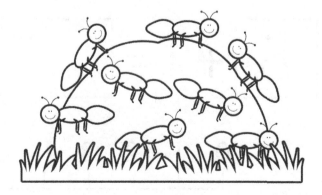

33

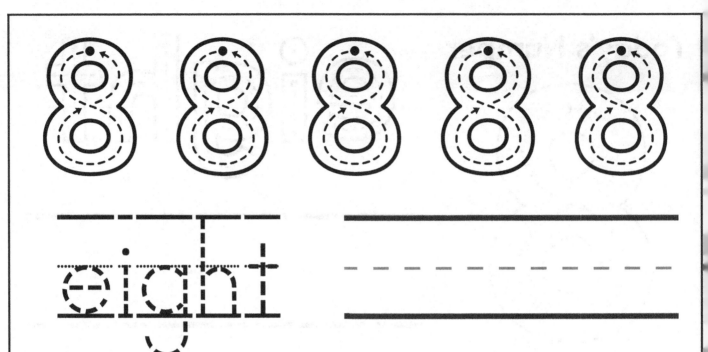

eight

---

 Draw 8 seeds.

 Draw 8 raindrops.

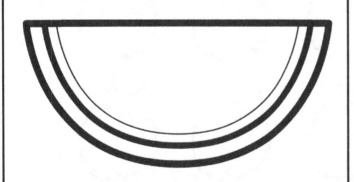

Add one.

Take away one.

$$8 + 1 = \boxed{\phantom{0}}$$

$$8 - 1 = \boxed{\phantom{0}}$$

0  1  2  3  4  5  6  7  8  9  10

Count and color 8 bears.

Find and color eight.

| 1 | 2 | 3 | 4 | 5 | 6 | 7 | 8 | 9 | 10 |

Draw 8 ornaments.

Draw 8 cookies.

I less                    I more

| | 8 | |

0  1  2  3  4  5  6  7  8  9  10

Make eight dots.

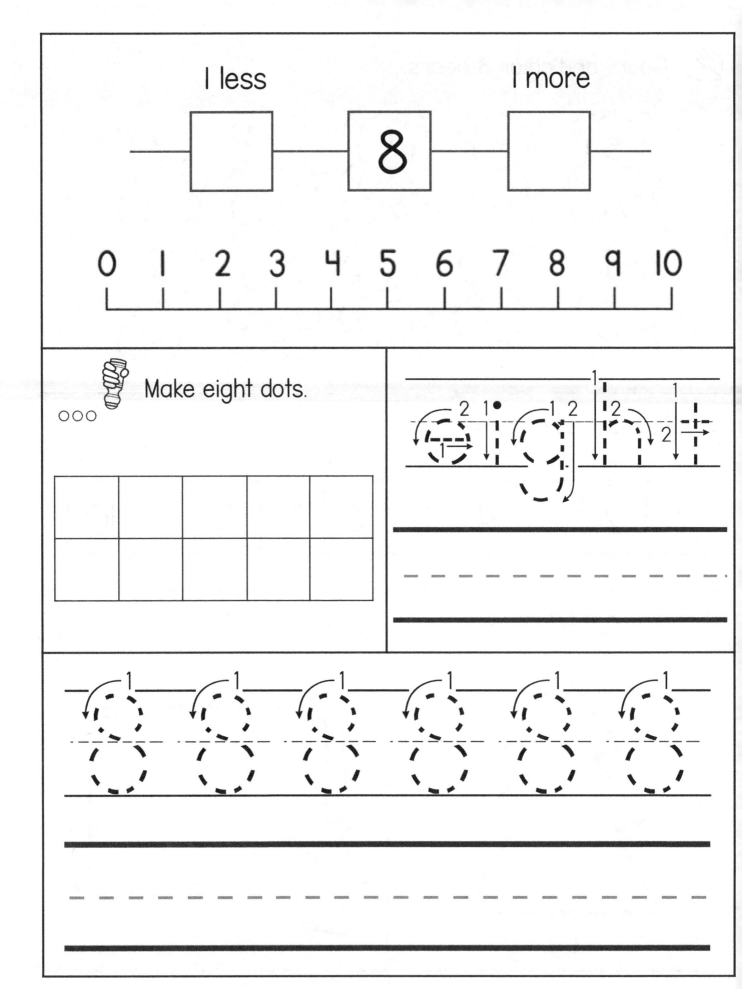

eight

# Today's Number

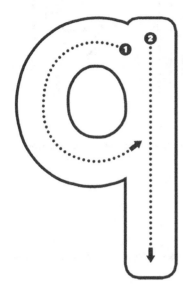

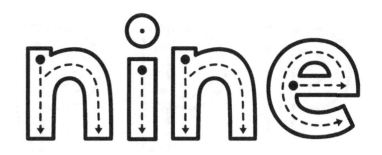

 Count and color 9 apples.

 Which has 9 butterflies?

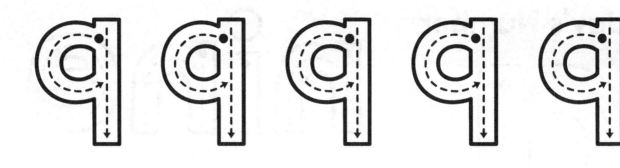

## Draw 9 sprinkles.

## Draw 9 sun beams.

Add one.

Take away one.

$9 + 1 = \boxed{\phantom{0}}$

$9 - 1 = \boxed{\phantom{0}}$

0  1  2  3  4  5  6  7  8  9  10

# Count and color 9 bears.

# Find and color nine.

| 1 | 2 | 3 | 4 | 5 | 6 | 7 | 8 | 9 | 10 |

# Draw 9 meatballs.

# Draw 9 teeth.

I less            I more

q

0   1   2   3   4   5   6   7   8   9   10

Make nine dots.

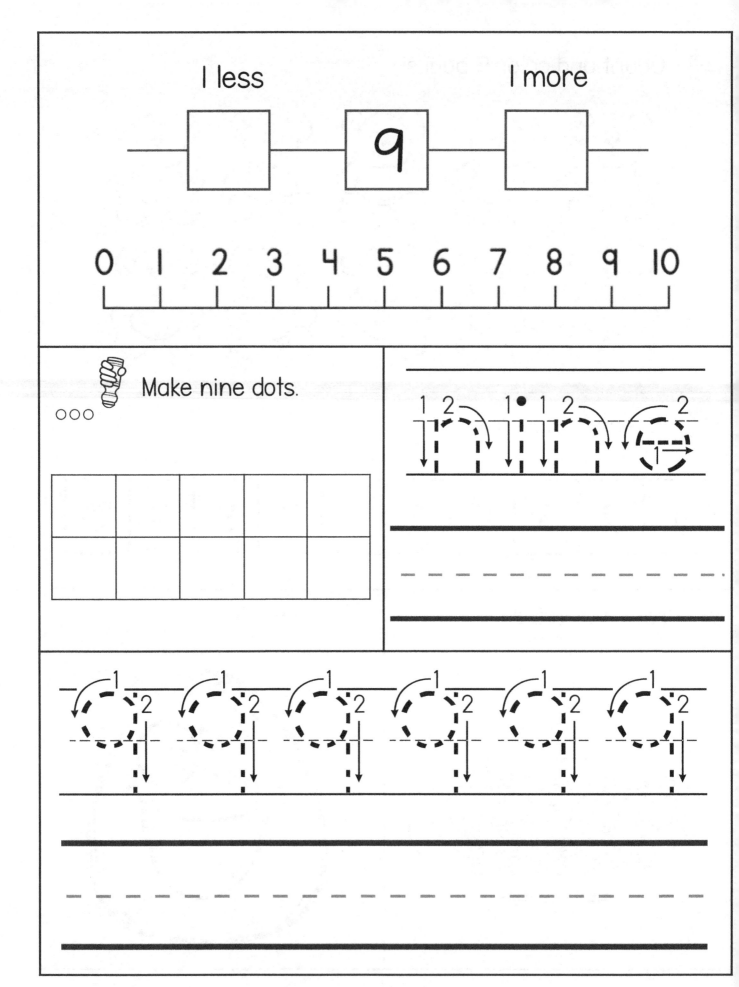

# Today's Number

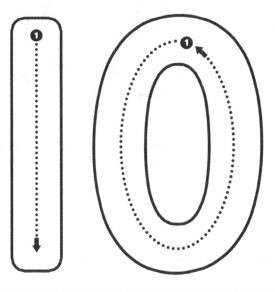
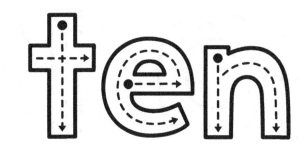

_____

Count and color 10 apples.

Which has 10 cookies?

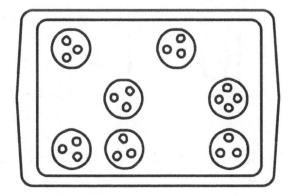
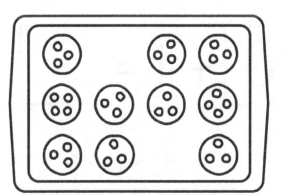

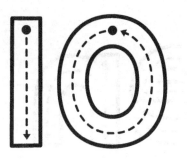  

ten

 Draw 10 legs.

 Draw 10 snowflakes.

Add one.

$\boxed{10} + 1 = \boxed{\phantom{0}}$

Take away one.

$\boxed{10} - 1 = \boxed{\phantom{0}}$

0 1 2 3 4 5 6 7 8 9 10 11 12 13 14 15 16 17 18 19 20

Count and color 10 bears.

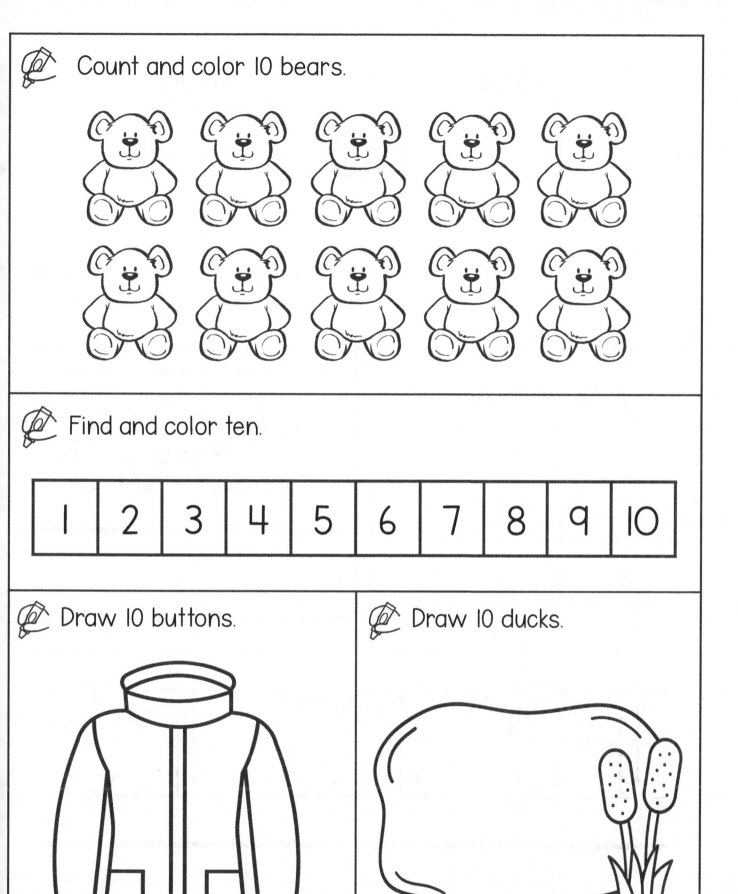

Find and color ten.

| 1 | 2 | 3 | 4 | 5 | 6 | 7 | 8 | 9 | 10 |

Draw 10 buttons.

Draw 10 ducks.

I less                    I more

| | 10 | |

0  1  2  3  4  5  6  7  8  9  10  11  12  13  14  15  16  17  18  19  20

Make ten dots.

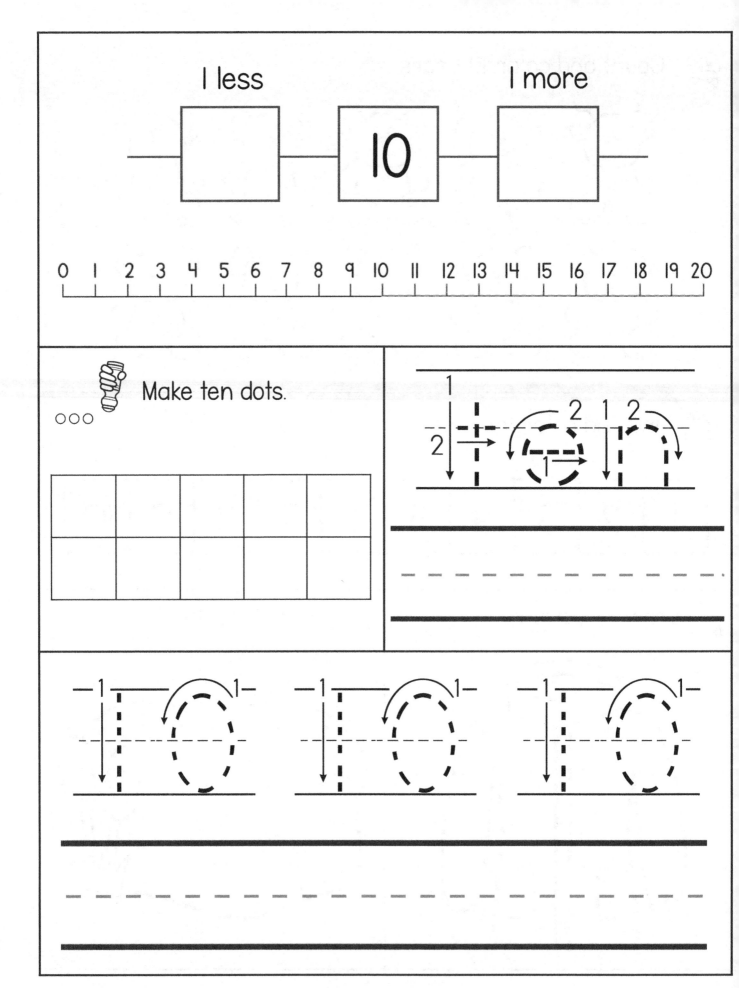

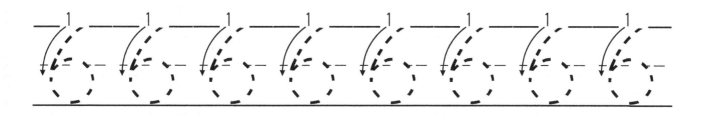

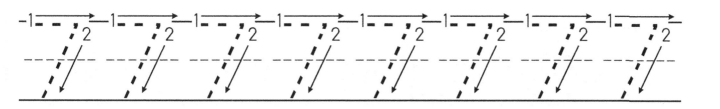

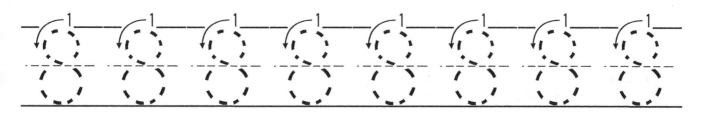

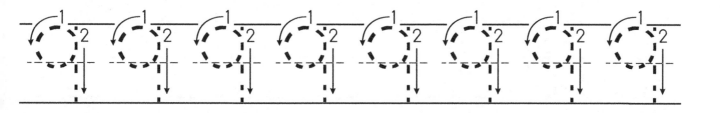

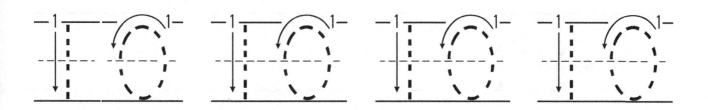

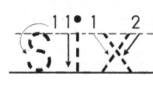

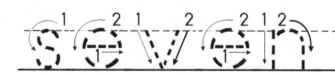

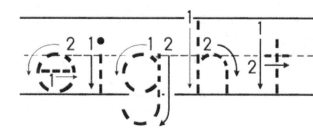

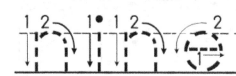

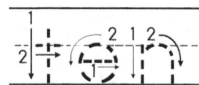

# Today's Number

## eleven

---

Make 11.

10 + ☐

---

Which has 11 gum balls?

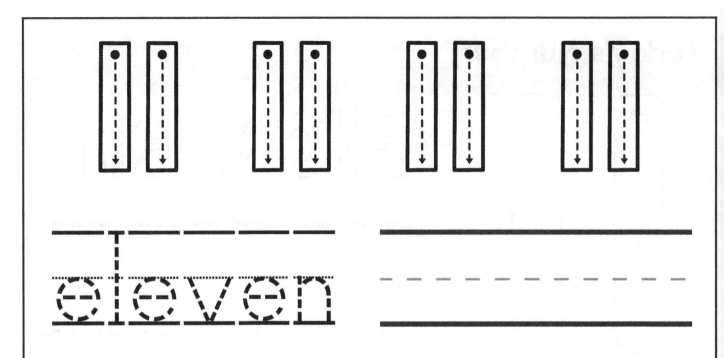

 Draw 11 apples.

 Draw 11 ants.

Add one.

Take away one.

$$\boxed{11} + 1 = \boxed{\phantom{0}}$$

$$\boxed{11} - 1 = \boxed{\phantom{0}}$$

0  1  2  3  4  5  6  7  8  9  10  11  12  13  14  15  16  17  18  19  20

✎ Count and color 11 bugs.

✎ Find and color eleven.

| 10 | 11 | 12 | 13 | 14 | 15 | 16 | 17 | 18 | 19 | 20 |

✎ Draw 11 sprinkles.

✎ Draw 11 ice cubes.

I less          I more

|         |   **II**   |         |

0  1  2  3  4  5  6  7  8  9  10  11  12  13  14  15  16  17  18  19  20

Is it 11?

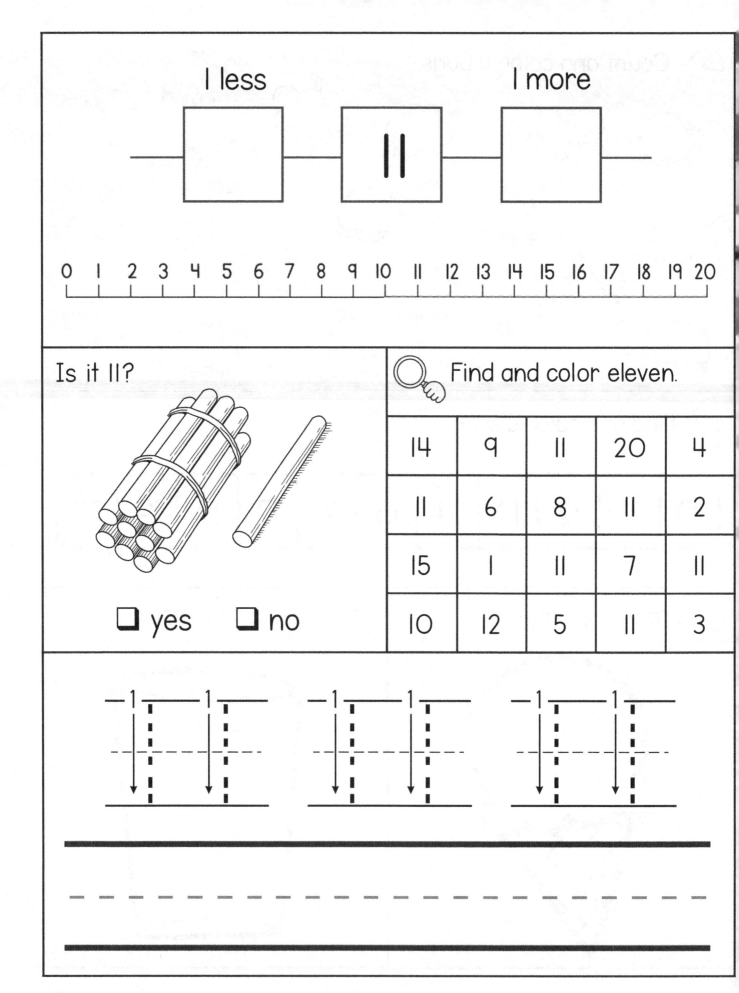

☐ yes    ☐ no

🔍 Find and color eleven.

| 14 | 9  | 11 | 20 | 4  |
|----|----|----|----|----|
| 11 | 6  | 8  | 11 | 2  |
| 15 | 1  | 11 | 7  | 11 |
| 10 | 12 | 5  | 11 | 3  |

# Today's Number

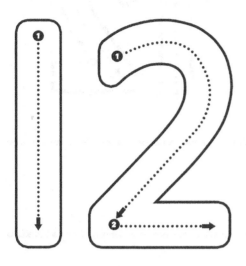

---

 Make 12.

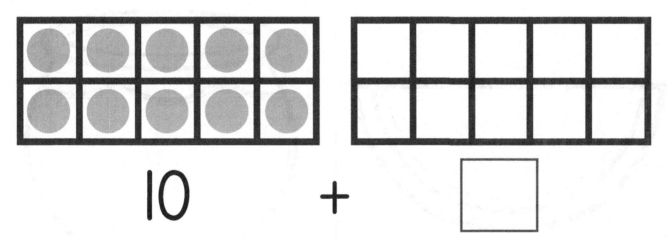

10  +  ☐

---

 Which has 12 cookies?

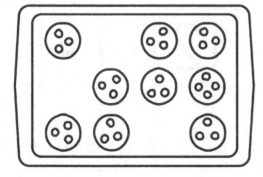

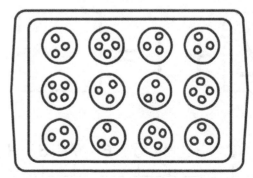

twelve

 Draw 12 seeds.

 Draw 12 fish.

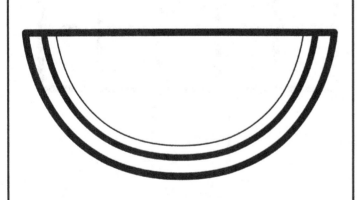

Add one.

Take away one.

$$12 + 1 = \boxed{\phantom{0}}$$

$$12 - 1 = \boxed{\phantom{0}}$$

0  1  2  3  4  5  6  7  8  9  10  11  12  13  14  15  16  17  18  19  20

✏️ Count and color 12 bugs.

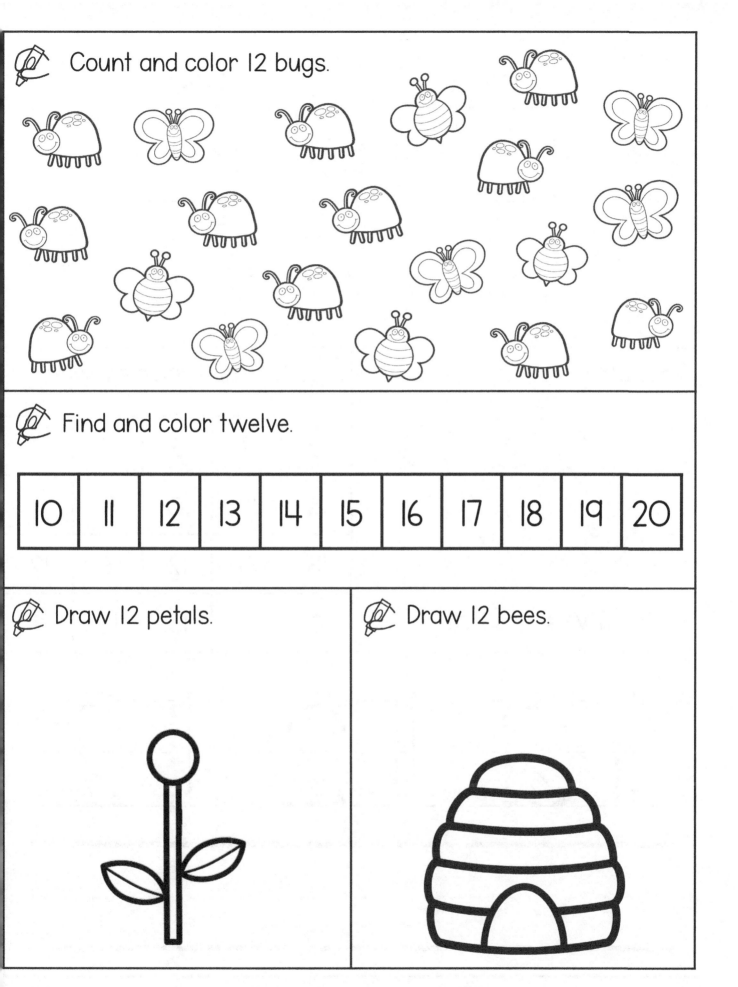

✏️ Find and color twelve.

| 10 | 11 | 12 | 13 | 14 | 15 | 16 | 17 | 18 | 19 | 20 |
|----|----|----|----|----|----|----|----|----|----|----|

✏️ Draw 12 petals.

✏️ Draw 12 bees.

I less                             I more

12

0   1   2   3   4   5   6   7   8   9   10   11   12   13   14   15   16   17   18   19   20

Is it 12?

☐ yes     ☐ no

Find and color twelve.

| 12 | 7 | 5 | 12 | 0 |
|----|----|----|----|----|
| 16 | 12 | 10 | 9 | 6 |
| 18 | 14 | 12 | 3 | 12 |
| 2 | 3 | 13 | 12 | 11 |

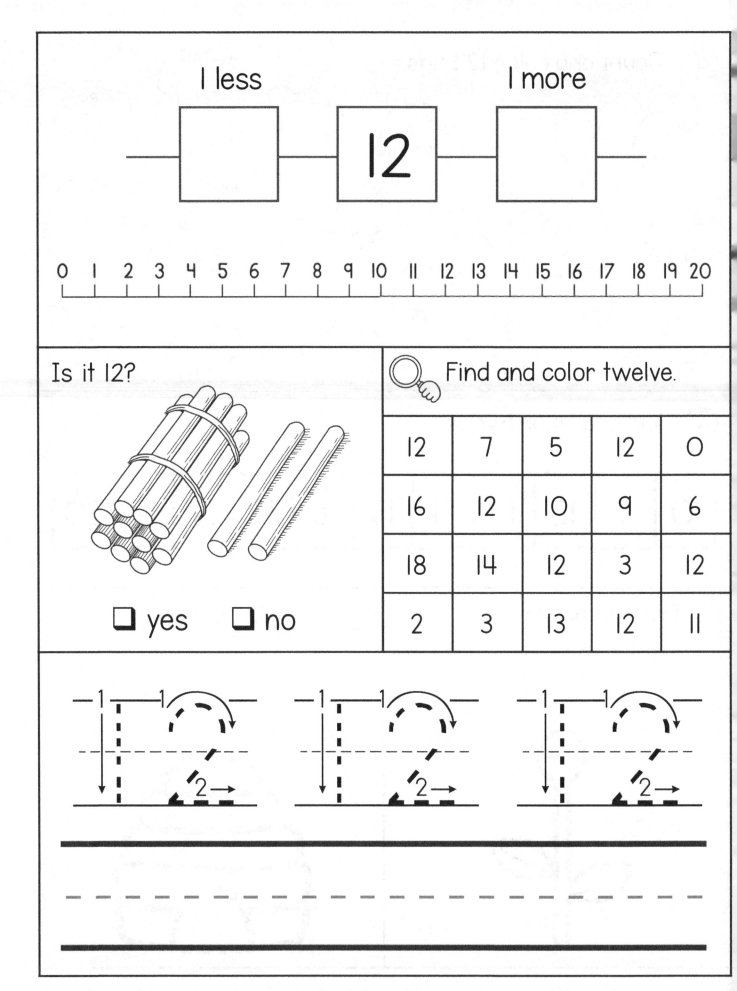

# Today's Number

13 thirteen

_____

- - - - - - - - - - - -

_____

✎ Make 13.

10  +  ☐

🔍 Which has 13 French fries?

thirteen _____

 Draw 13 hairs.　　　　 Draw 13 candies.

Add one.　　　　　　　Take away one.

13 + 1 = ☐　　　　　13 - 1 = ☐

0  1  2  3  4  5  6  7  8  9  10  11  12  13  14  15  16  17  18  19  20

# Count and color 13 bugs.

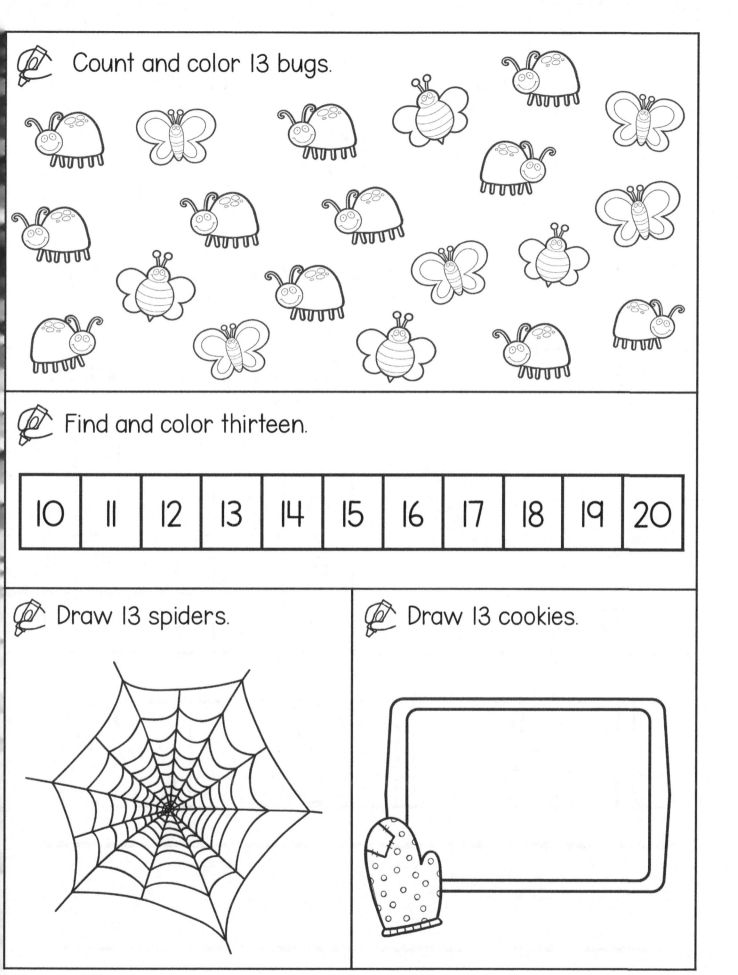

# Find and color thirteen.

| 10 | 11 | 12 | 13 | 14 | 15 | 16 | 17 | 18 | 19 | 20 |
|----|----|----|----|----|----|----|----|----|----|----|

# Draw 13 spiders.

# Draw 13 cookies.

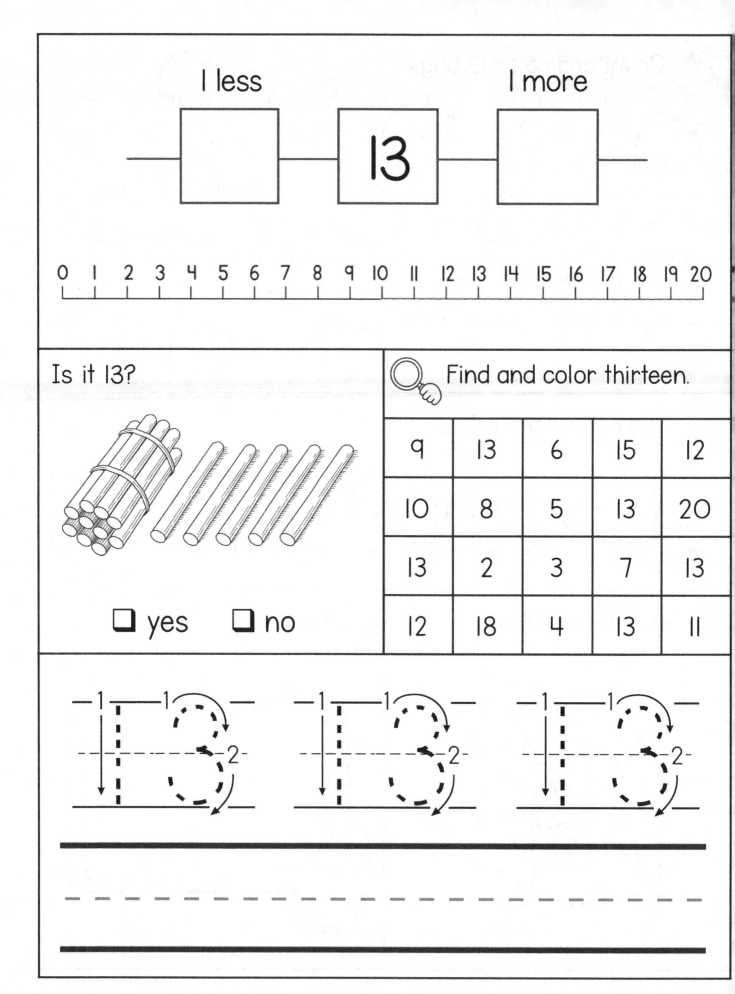

I less                I more

13

0 1 2 3 4 5 6 7 8 9 10 11 12 13 14 15 16 17 18 19 20

Is it 13?

☐ yes    ☐ no

Find and color thirteen.

| 9 | 13 | 6 | 15 | 12 |
|---|---|---|---|---|
| 10 | 8 | 5 | 13 | 20 |
| 13 | 2 | 3 | 7 | 13 |
| 12 | 18 | 4 | 13 | 11 |

# Today's Number

14  fourteen

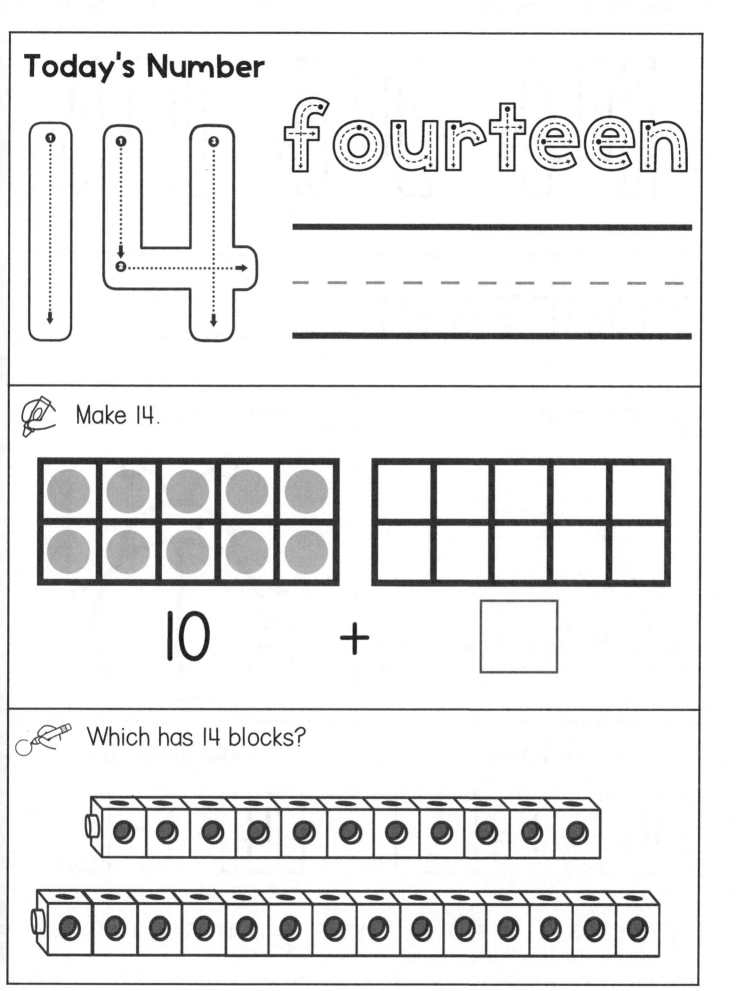

Make 14.

10  +  [ ]

Which has 14 blocks?

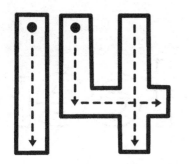 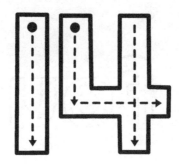 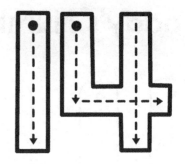

fourteen _____

 Draw 14 ants.

Draw 14 spots.

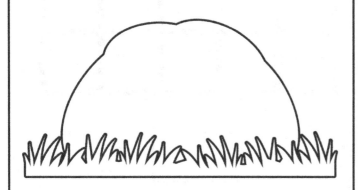

Add one.

14 + 1 = ☐

Take away one.

14 - 1 = ☐

0  1  2  3  4  5  6  7  8  9  10  11  12  13  14  15  16  17  18  19  20

Count and color 14 bugs.

Find and color fourteen.

| 10 | 11 | 12 | 13 | 14 | 15 | 16 | 17 | 18 | 19 | 20 |

Draw 14 pieces of popcorn.

Draw 14 water drops.

I less                         I more

14

0 1 2 3 4 5 6 7 8 9 10 11 12 13 14 15 16 17 18 19 20

Is it 14?

☐ yes     ☐ no

Find and color fourteen.

| 2 | 17 | 14 | 12 | 9 |
|---|----|----|----|---|
| 10 | 14 | 6 | 4 | 18 |
| 13 | 5 | 12 | 14 | 16 |
| 14 | 10 | 14 | 8 | 19 |

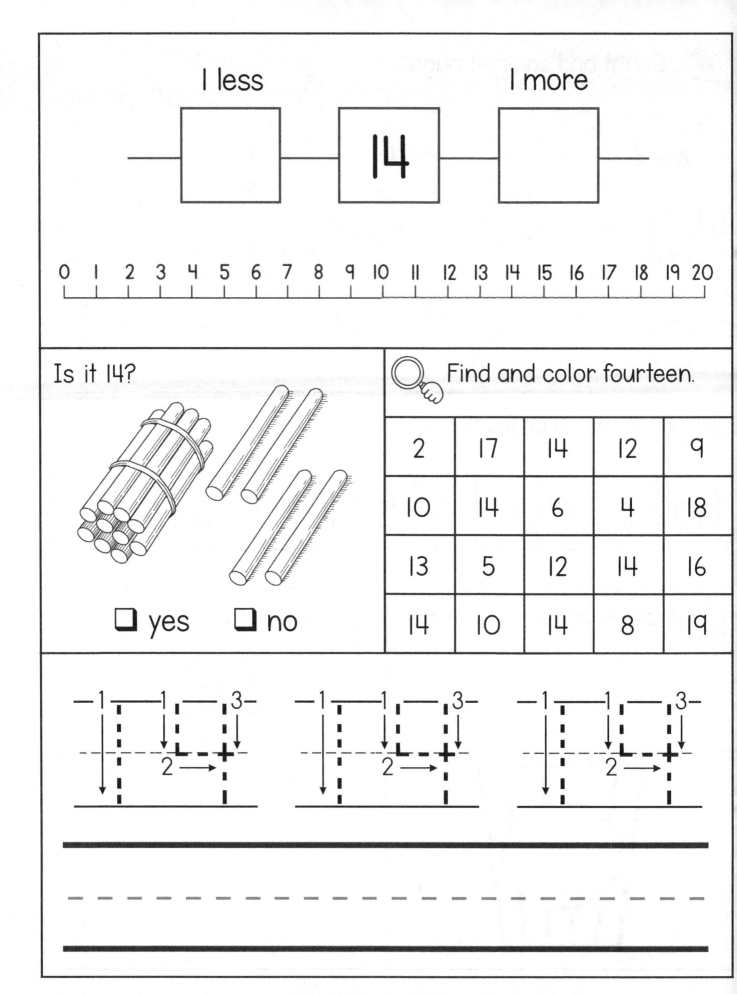

# Today's Number

15    fifteen

_____
- - - - - - - - - - - - - -
_____

✎ Make 15.

10   +   ☐

🔍 Which has 15 snowballs?

fifteen _____

 Draw 15 chocolate chips.

 Draw 15 bubbles.

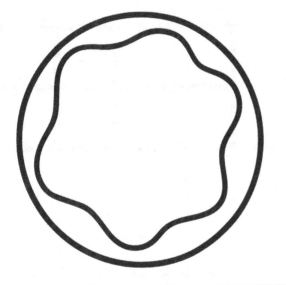

Add one.

15 + 1 = ☐

Take away one.

15 - 1 = ☐

0  1  2  3  4  5  6  7  8  9  10  11  12  13  14  15  16  17  18  19  20

Count and color 15 bugs.

Find and color fifteen.

| 10 | 11 | 12 | 13 | 14 | 15 | 16 | 17 | 18 | 19 | 20 |
|----|----|----|----|----|----|----|----|----|----|----|

Draw 15 teeth.

Draw 15 shells.

65

I less                    I more

|     | 15 |     |

0  1  2  3  4  5  6  7  8  9  10  11  12  13  14  15  16  17  18  19  20

Is it 15?

☐ yes    ☐ no

Find and color fifteen.

| 2 | 13 | 8 | 15 | 7 |
|---|----|---|----|---|
| 14 | 15 | 20 | 9 | 10 |
| 5 | 12 | 6 | 15 | 18 |
| 15 | 4 | 17 | 16 | 15 |

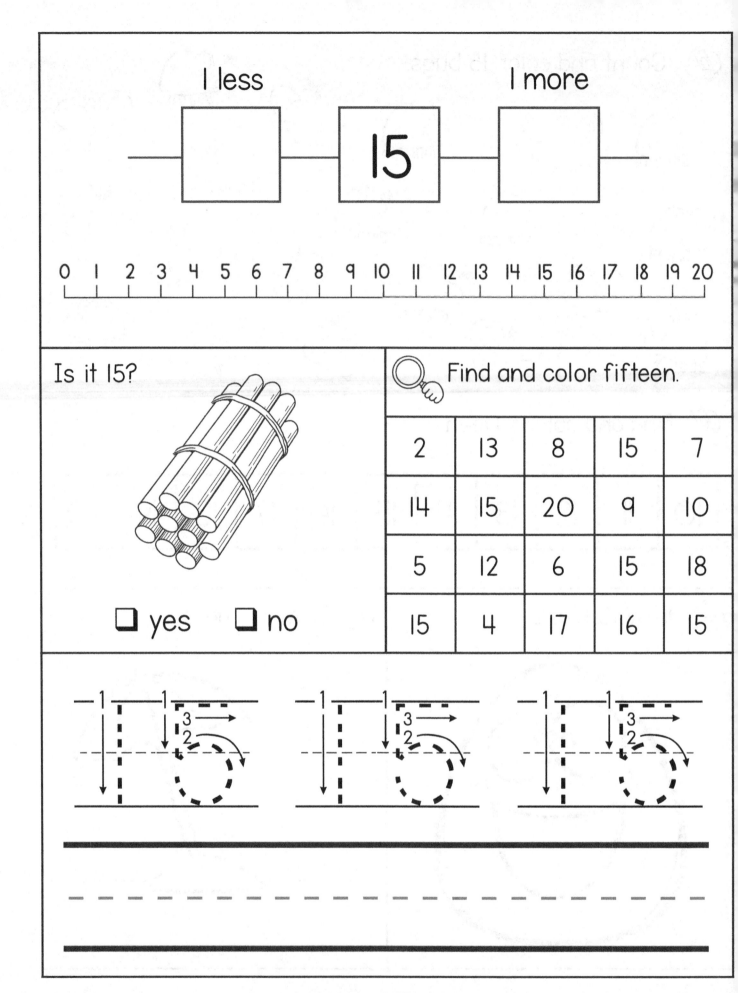

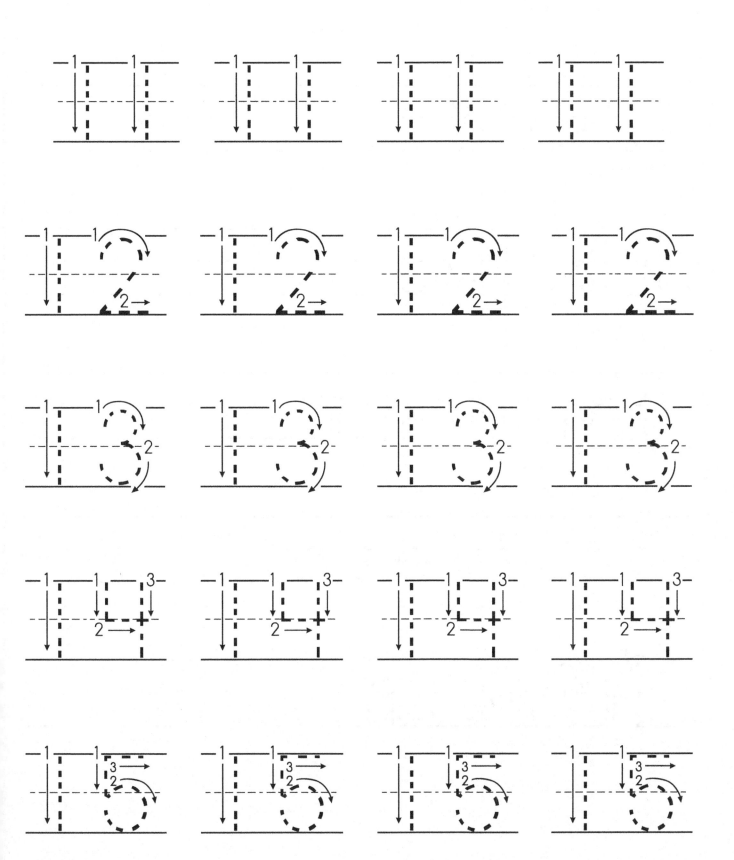

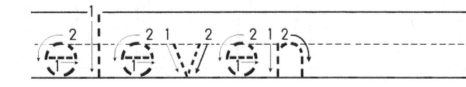

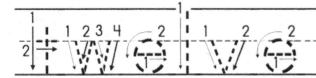

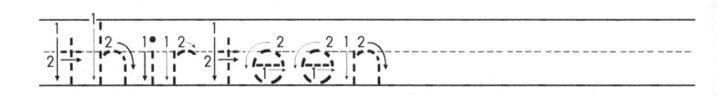

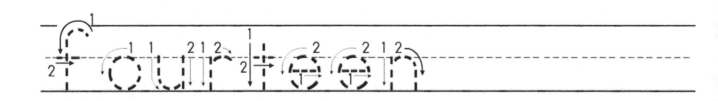

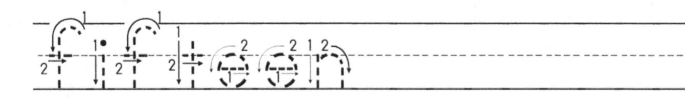

# Today's Number

16 sixteen

## Make 16.

10 + [ ]

## Which has 16 beads?

 Draw 16 candles.

 Draw 16 legs.

Add one.

Take away one.

16 + 1 = ☐

16 - 1 = ☐

0  1  2  3  4  5  6  7  8  9  10  11  12  13  14  15  16  17  18  19  20

Count and color 16 bugs.

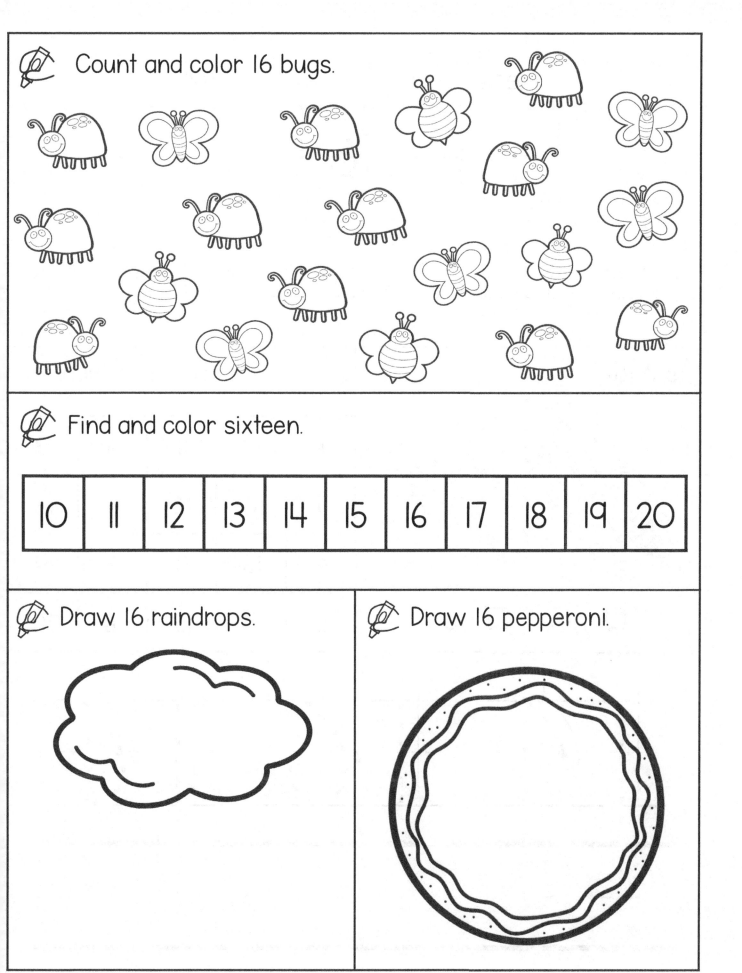

Find and color sixteen.

| 10 | 11 | 12 | 13 | 14 | 15 | 16 | 17 | 18 | 19 | 20 |
|----|----|----|----|----|----|----|----|----|----|----|

Draw 16 raindrops.

Draw 16 pepperoni.

I less                    I more

| | 16 | |

0  1  2  3  4  5  6  7  8  9  10  11  12  13  14  15  16  17  18  19  20

Is it 16?

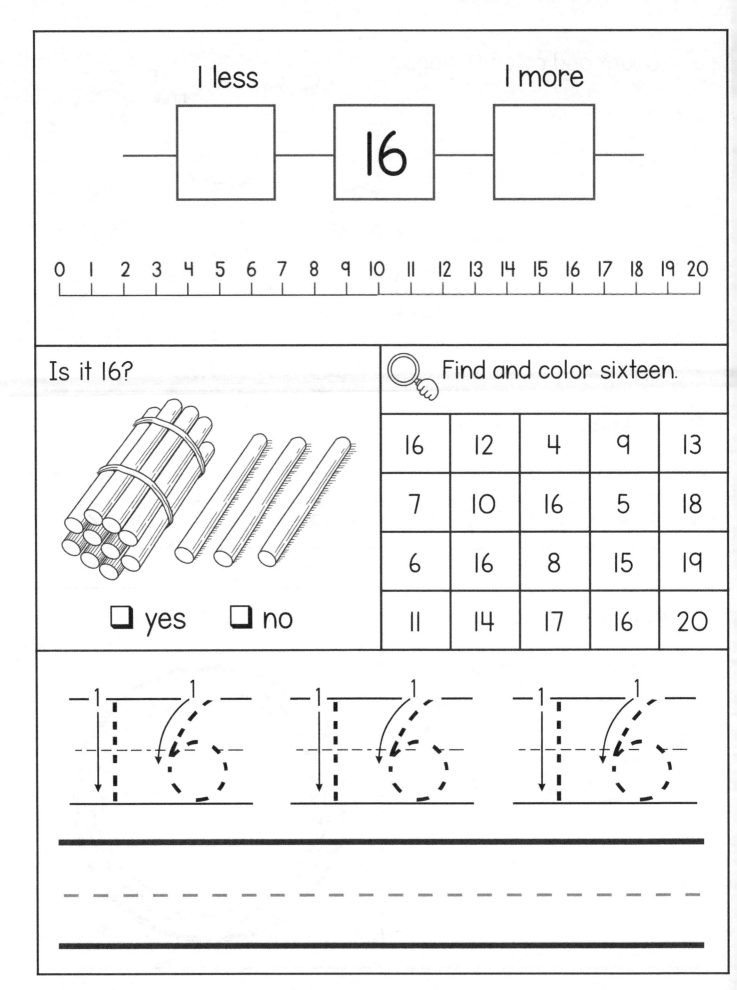

☐ yes      ☐ no

🔍 Find and color sixteen.

| 16 | 12 | 4 | 9 | 13 |
|----|----|----|----|----|
| 7 | 10 | 16 | 5 | 18 |
| 6 | 16 | 8 | 15 | 19 |
| 11 | 14 | 17 | 16 | 20 |

# Today's Number

17 seventeen

---

## Make 17.

$$10 \quad + \quad \boxed{\phantom{0}}$$

---

## Which has 17 gumballs?

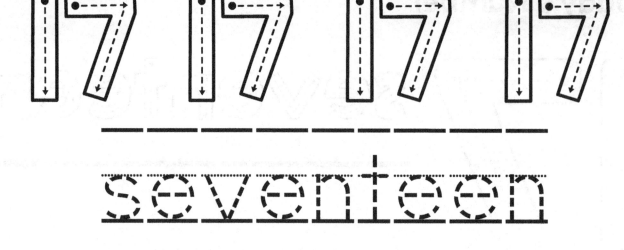

_____

seventeen

_____

_ _ _ _ _ _ _ _

_____

✏ Draw 17 spots.

✏ Draw 17 seeds.

Add one.

Take away one.

17 + 1 = ☐

17 - 1 = ☐

0  1  2  3  4  5  6  7  8  9  10  11  12  13  14  15  16  17  18  19  20

Count and color 17 bugs.

Find and color seventeen.

| 10 | 11 | 12 | 13 | 14 | 15 | 16 | 17 | 18 | 19 | 20 |

Draw 17 stars.

Draw 17 ornaments.

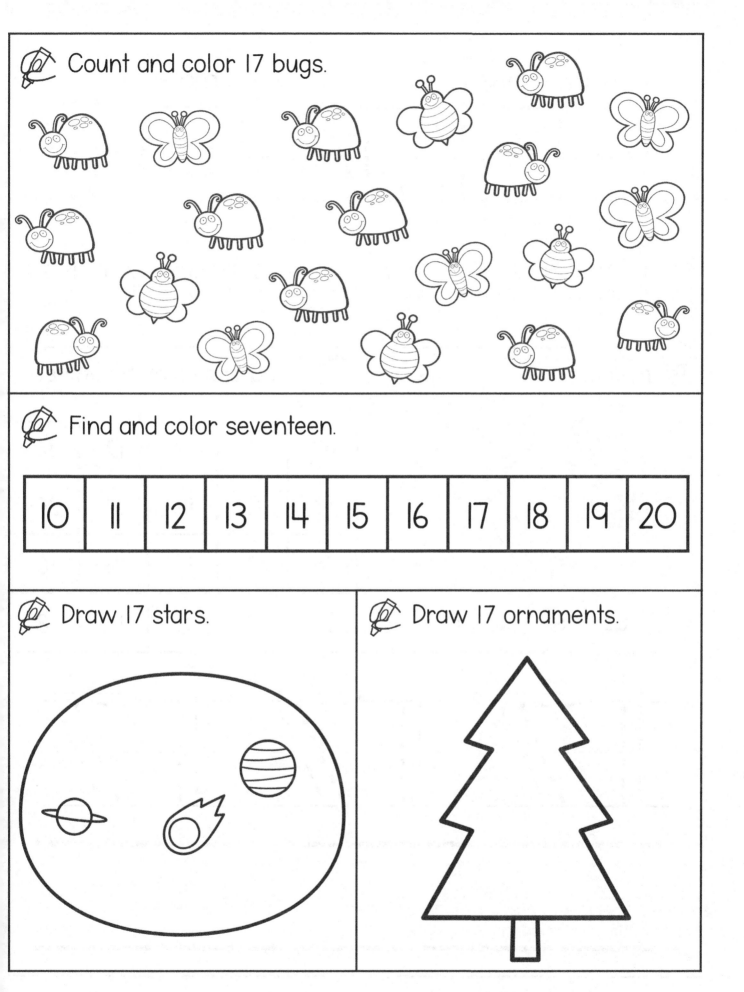

I less               I more

17

0  1  2  3  4  5  6  7  8  9  10  11  12  13  14  15  16  17  18  19  20

Is it 17?

☐ yes ☐ no

Find and color seventeen.

| 10 | 17 | 5 | 12 | 9 |
| 13 | 4 | 19 | 17 | 20 |
| 7 | 2 | 17 | 14 | 16 |
| 17 | 3 | 8 | 20 | 18 |

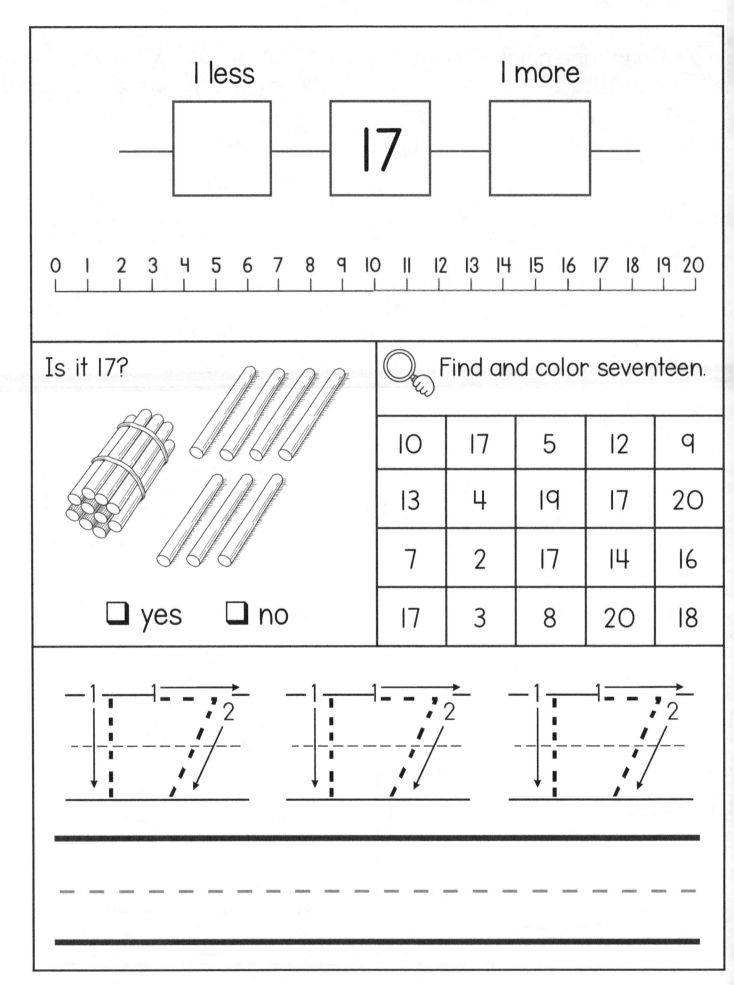

# Today's Number

18  eighteen

Make 18.

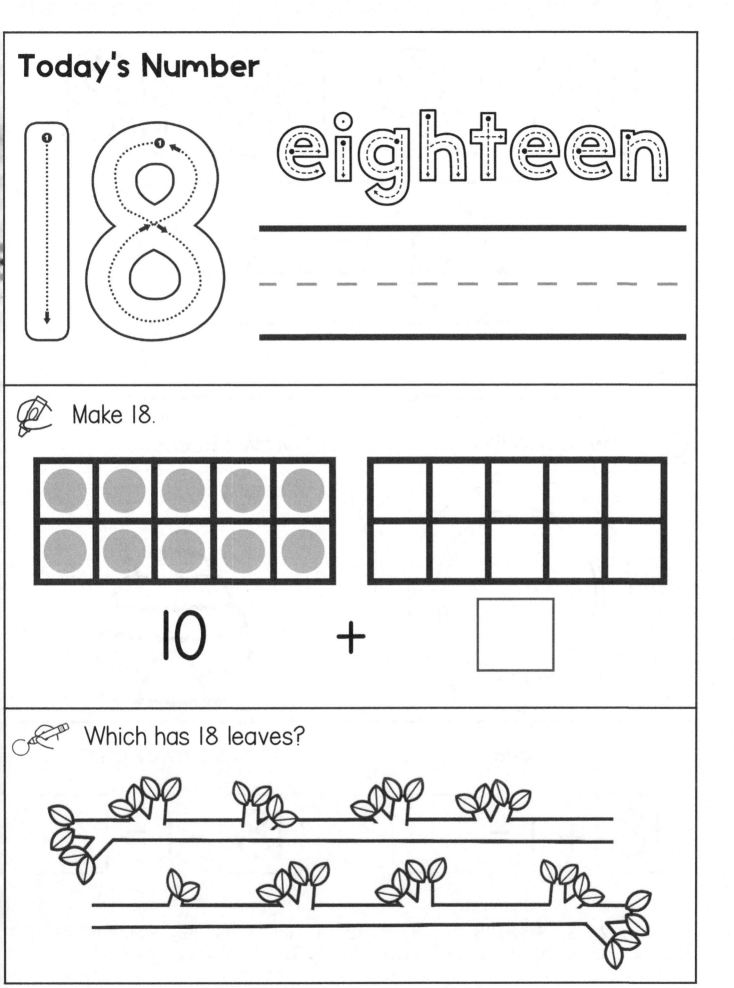

10  +  ☐

Which has 18 leaves?

eighteen

Draw 18 noodles.

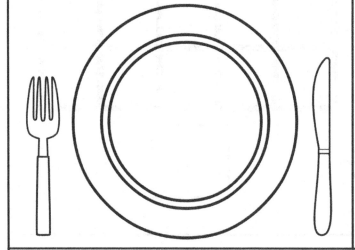

Draw 18 bees.

Add one.

18 + 1 = ☐

Take away one.

18 - 1 = ☐

| 0 | 1 | 2 | 3 | 4 | 5 | 6 | 7 | 8 | 9 | 10 | 11 | 12 | 13 | 14 | 15 | 16 | 17 | 18 | 19 | 20 |

Count and color 18 bugs.

Find and color eighteen.

| 10 | 11 | 12 | 13 | 14 | 15 | 16 | 17 | 18 | 19 | 20 |
|----|----|----|----|----|----|----|----|----|----|----|

Draw 18 sun beams.

Draw 18 apples.

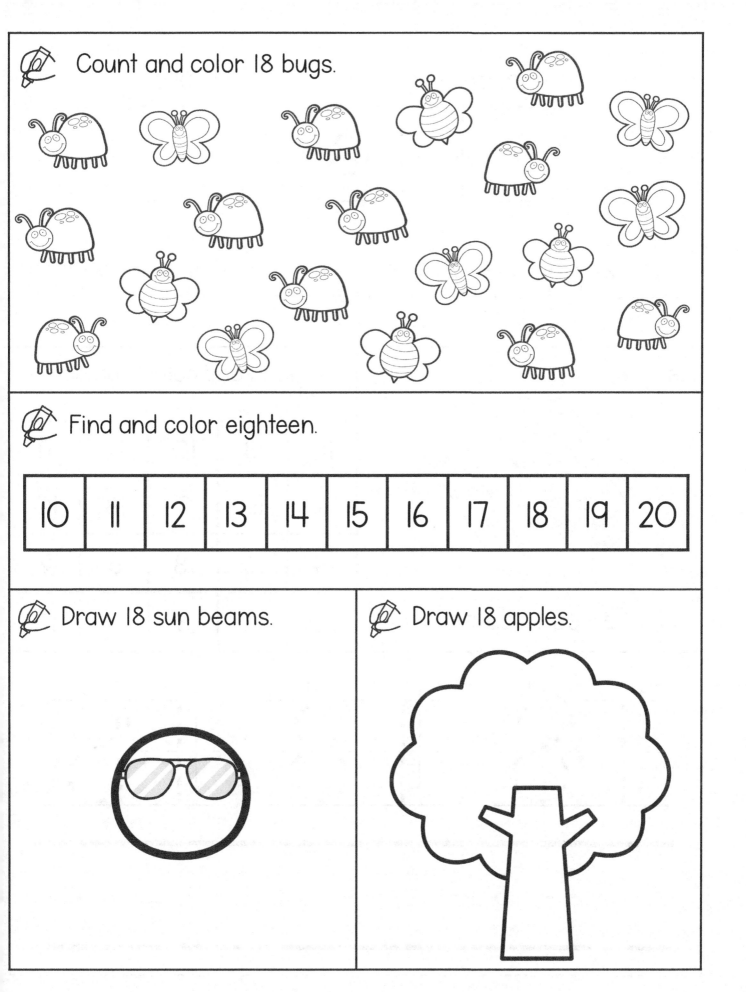

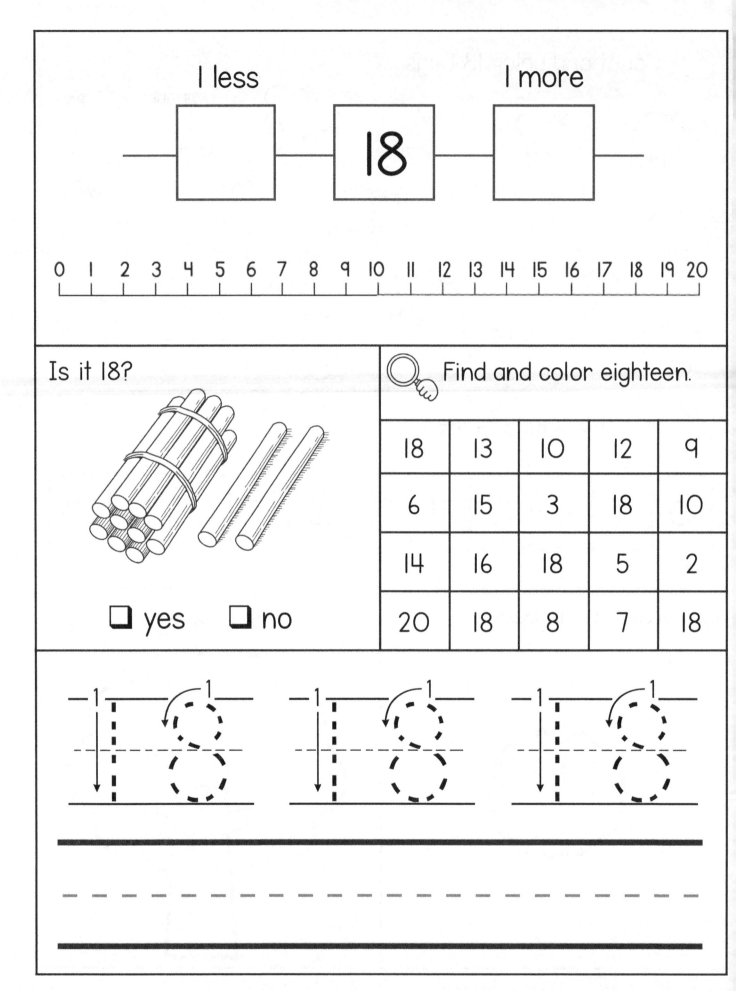

1 less

1 more

18

0 1 2 3 4 5 6 7 8 9 10 11 12 13 14 15 16 17 18 19 20

Is it 18?

☐ yes    ☐ no

Find and color eighteen.

| 18 | 13 | 10 | 12 | 9 |
|----|----|----|----|----|
| 6 | 15 | 3 | 18 | 10 |
| 14 | 16 | 18 | 5 | 2 |
| 20 | 18 | 8 | 7 | 18 |

# Today's Number

19 nineteen

---

✏️ Make 19.

10 + ☐

---

🔍 Which has 19 beads?

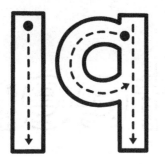 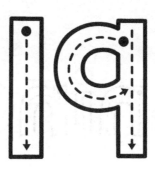 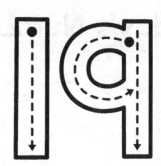

nineteen

 Draw 19 sprinkles.

 Draw 19 nuts.

Add one.

Take away one.

$19 + 1 = \boxed{\phantom{00}}$

$19 - 1 = \boxed{\phantom{00}}$

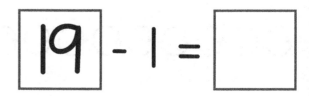

0 1 2 3 4 5 6 7 8 9 10 11 12 13 14 15 16 17 18 19 20

Count and color 19 bugs.

Find and color nineteen.

| 10 | 11 | 12 | 13 | 14 | 15 | 16 | 17 | 18 | 19 | 20 |
|---|---|---|---|---|---|---|---|---|---|---|

Draw 19 bugs

Draw 19 snowflakes.

I less                I more

[ ] | **19** | [ ]

0  1  2  3  4  5  6  7  8  9  10  11  12  13  14  15  16  17  18  19  20

Is it 19?

☐ yes    ☐ no

Find and color nineteen.

| 10 | 19 | 16 | 3 | 8 |
|----|----|----|----|----|
| 19 | 12 | 17 | 6 | 19 |
| 20 | 15 | 4 | 18 | 9 |
| 7 | 20 | 5 | 19 | 11 |

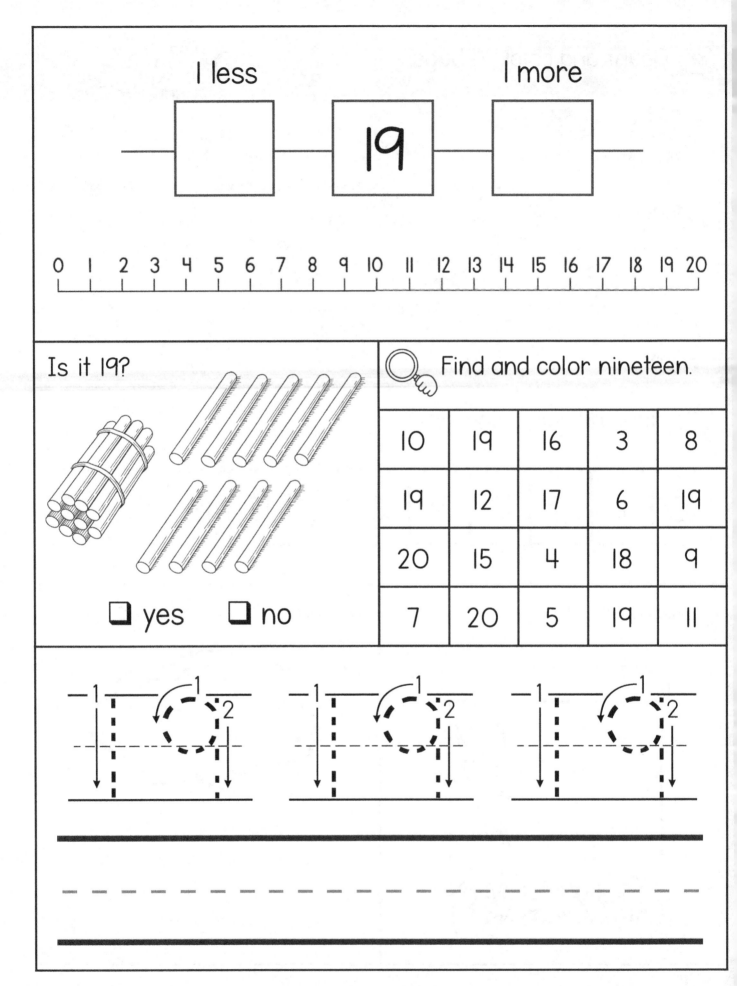

# Today's Number

**20** **twenty**

_____
- - - - - - - - - - -
_____

✎ Make 20.

10 + ☐

✎ Which has 20 gum balls?

twenty _____

 Draw 20 seeds.

 Draw 20 hairs.

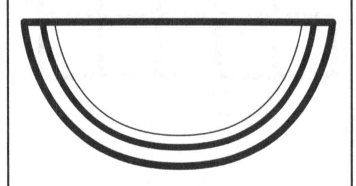

Add one.

Take away one.

20 + 1 = ☐

20 - 1 = ☐

0  1  2  3  4  5  6  7  8  9  10  11  12  13  14  15  16  17  18  19  20

# Count and color 20 bugs.

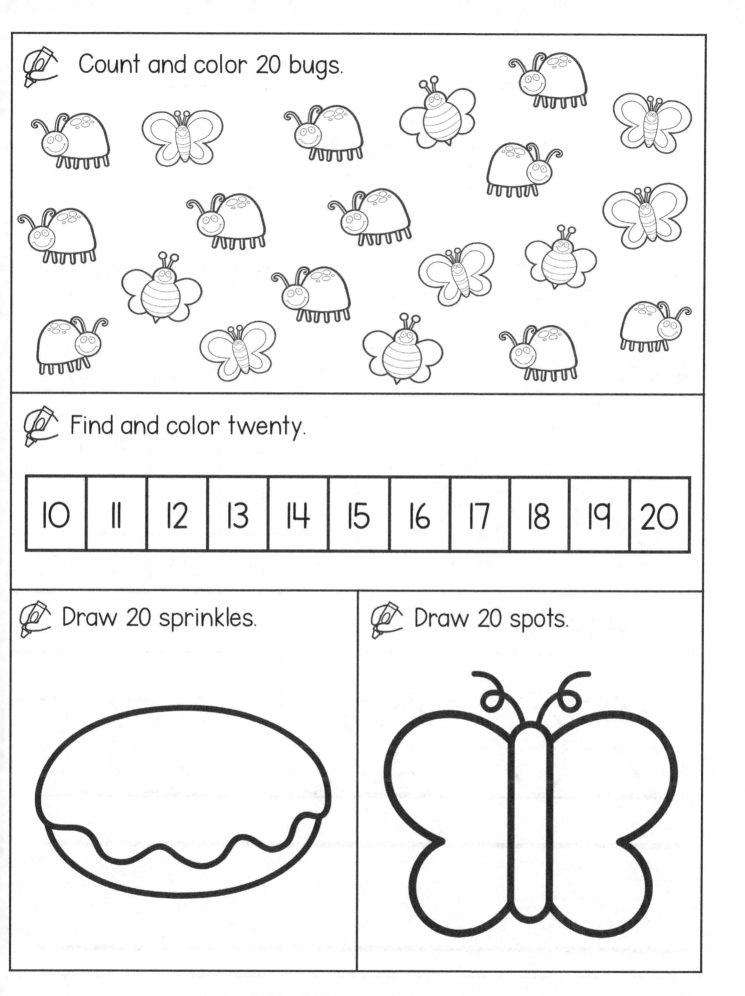

# Find and color twenty.

| 10 | 11 | 12 | 13 | 14 | 15 | 16 | 17 | 18 | 19 | 20 |
|----|----|----|----|----|----|----|----|----|----|----|

# Draw 20 sprinkles.

# Draw 20 spots.

I less                    I more

| | 20 | |

0  1  2  3  4  5  6  7  8  9  10  11  12  13  14  15  16  17  18  19  20

Is it 20?

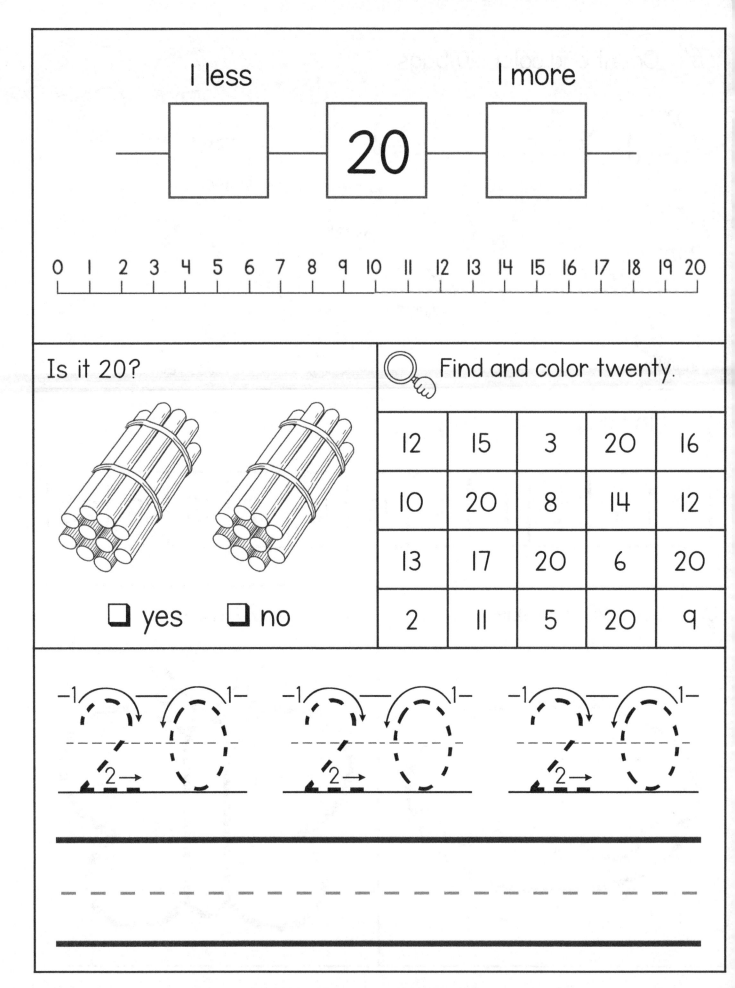

☐ yes    ☐ no

Find and color twenty.

| 12 | 15 | 3 | 20 | 16 |
| 10 | 20 | 8 | 14 | 12 |
| 13 | 17 | 20 | 6 | 20 |
| 2 | 11 | 5 | 20 | 9 |

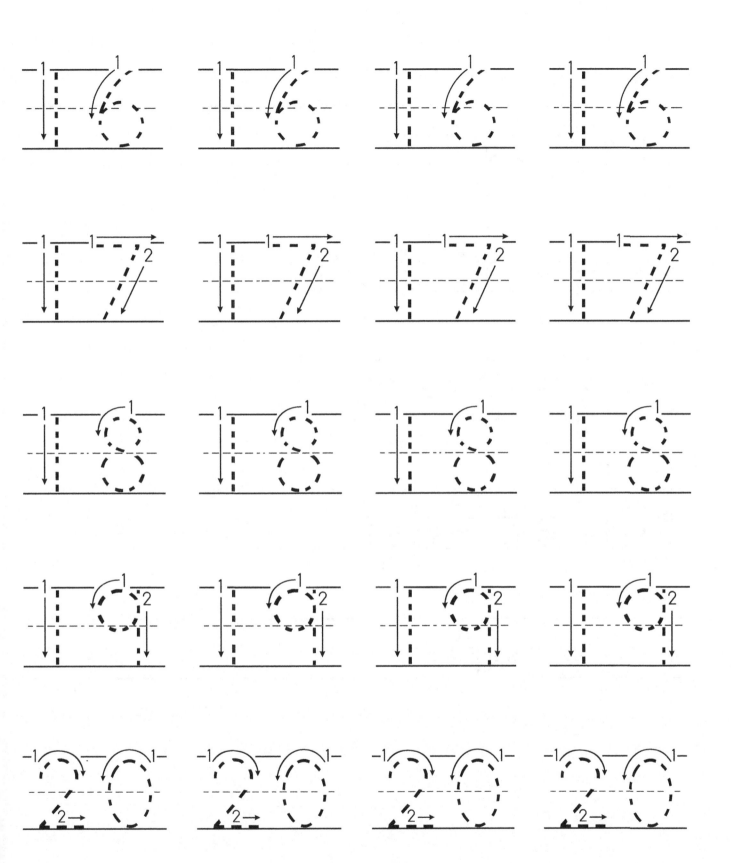

sixteen

seventeen

eighteen

nineteen

twenty

Made in the USA
Monee, IL
12 January 2025

76681260R00050